AF808319

Chocolate.
A Quick Immersion

Quick Immersions provide illuminating introductions to diverse topics in the worlds of social science, the hard sciences, philosophy and humanities. Written in clear and straightforward language by prestigious authors, the texts also offer valuable insights to readers seeking a deeper knowledge of those fields.

Claudi Mans

CHOCOLATE

A Quick Immersion

Tibidabo Publishing
New York

Published by Tibidabo Publishing, Inc. New York.

Copyediting by Lori Gerson
Cover art by Raimon Guirado

First published: April 2026

Visit our Series on our Web:
www.quickimmersions.com

ISBN: 978-1-949845-42-6
1 2 3 4 5 6 7 8 9 10

Library of Congress Control Number: 2026930296

Printed in the United States of America.

This work was published with the help of a grant from the Institut Ramon Llull.

Translated from Catalan by Lori Gerson.

institut
ramon llull

Contents

The Paradigm by Oriol Balaguer

Catalan pastry chef and chocolatier **Oriol Balaguer** (1971) designed a chocolate cake in 1993 that won the award for best dessert in the world in Lyon (France) in 2001. With some small changes, he sells it in his stores with the name of *Chocolate Paradigm* or *8 Textures of Chocolate.*

An expert taster has written that the cake "consists of fusing eight different layers of chocolate to achieve a balance of flavors and shapes. [...] Inside there is chocolate mousse, chocolate sponge cake, a crunchy part, and the creaminess and softness of the rest of the ingredients." But are the mousse or the sponge cake textures? Are the crunchy part, creaminess, softness textures? Is each layer of chocolate, mousse, sponge cake an ingredient? Or rather are they components, while the ingredients are the milk, cocoa, sugar, butter?

Are there eight different textures of chocolate? Or, before that, are there eight different types of chocolate? This book is dedicated to answering these questions and, above all, to understanding the relationship between the textures and the composition of chocolate. And it is not easy, because defining what texture is – strictly speaking, the textures – requires a prior understanding of many concepts.

When talking about any scientific or culinary topic, we must be especially careful with the terminology: We must be just as precise as pastry chefs and chocolatiers are when they make their preparations. In their recipes they use scales that weigh ingredients to a fraction of a gram, they heat or cool to temperatures measured in degrees of Celsius with precision thermometers and they use chronometers to adjust the times to the second. If they didn't do it like this, they would not be able to make paradigmatic preparations... And we must make the effort to do the same when writing.

The Book's Structure

Food is the essential material for survival as individuals. But it also has – at least in the developed world – many other functions: it is a source of pleasure in the gastronomic preparations we eat, and it can also be consolation in difficult times when we take refuge in food, perhaps compulsively, and in drinking. And among the enormous variety of foods we have at our fingertips – in the developed world, let's remember – chocolate is probably the one that best combines the triple function of nourishing us, giving us pleasure and comforting us. That's why chocolate has taken over the planet, becoming one of the iconic and essential foods for many people.

But there is more. Chocolate is food with a large amount of calories, and this can create a feeling of guilt in consumers who do not want to get fat, especially if they are diabetic or afraid of getting cavities. Additionally, some of the components in cocoa, such as theobromine, have a component that can be, to a greater or lesser degree, addictive, and also a euphoric component. When consuming chocolate in any of its forms we find pleasure, consolation, health problems, addiction and euphoria.

The aim of this book is to analyze, with scientific arguments, what chocolate is and why the same product creates such different sensations in us. To do so, we will follow the traditional structure of how to approach food. Following a classic division from the food sciences, we will see the food from tree to mouth, that is, what chocolate is (Chapter 1) and its composition (Chapter 2). To this point we are in the discipline of the science and technology of food. After this we put the product in our mouth, and that places us in the discipline of the sensory sciences or gastronomic sciences. These will be in Chapter 3 (textures) and Chapter 4 (other organoleptic sensations). If we ingest chocolate, that is, through the mouth to inside the body, we would be in the field of nutrition and dietetics, a topic we will dedicate a very short chapter to, number 5, with questions and answers.

In the end, we will have analyzed the relationship between chemical composition, physical and structural characterization and the texture and flavor of the product, which determines the sensory experience created by its consumption. And which we expect to be pleasant. Our hope is that the book will give you this feeling as well.

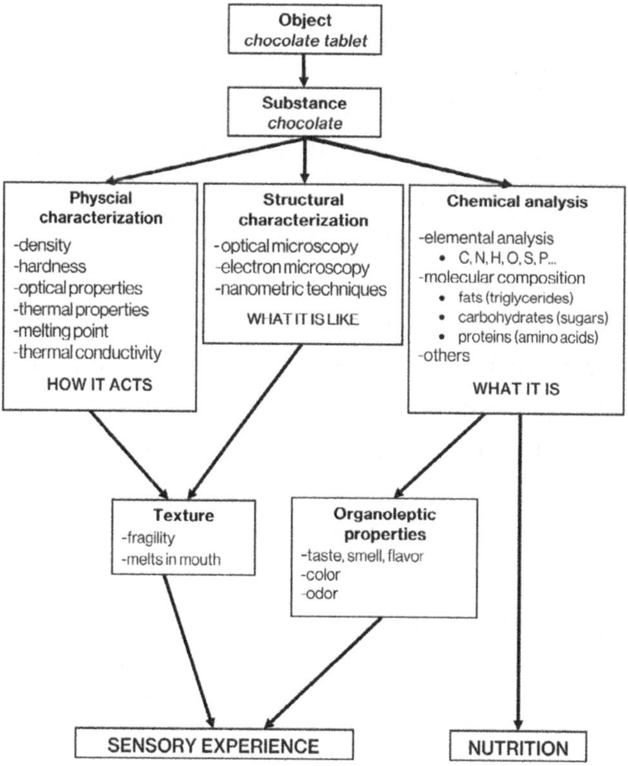

Figure 1. Book structure: progression of knowledge over the course of the text. Source: prepared by author.

Figure 1 shows in more detail the stages of the itinerary we will follow. We will not characterize objects made with chocolate – a tablet, a bunny, ice cream, a bonbon – but rather the substance itself. On the one hand, its simplest physical properties, such as density or hardness, and its thermal and optical properties, that is, color, transparency, shine. All this gives us information about *how the substance acts*. On the other hand, various techniques based on microscopy let us know about *what* the internal structure of the substance *is*: granules, crystals, drops, etc. All this tells us what it is like on the inside. And finally, different chemical techniques let us know *what* the substance *is*, what elements it is made of and, above all, what the molecules are that form it: fats such as triglycerides, sugars, carbohydrates, and proteins made up of amino acids are the main ones, and it can have many more, such as vitamins, food coloring or additives.

With these characteristics we can understand *why chocolate has the texture it has*, and why it has the characteristics of taste, smell, color and flavor that define it. We will be able to understand what the sensory experience when it is ingested is based on. And all this practically apart from, as has been noted, the nutritional aspects, which also depend on its chemical composition.

This structure-itinerary will gradually reveal various questions that often arise when we approach chocolate with a curious and investigative spirit.

Chapter 1

Chocolate. Its ingredients

Who invented the term "chocolate"?

It might seem an irrelevant question at first glance, but we need to start at the beginning. And that means knowing why things are called what they are called. As is well known, chocolate is not a natural product but rather a mixture that was invented in the Mayan civilization, long before Spanish colonization in the 15th century. The priests and warriors there mixed previously fermented cocoa beans with different local spices such as *achiote* (red vegetable food coloring), vanilla and pepper. They ground it and heated it, and the melted cocoa butter was removed from the paste. The resulting product was allowed to dry and the remaining solid was dissolved with hot water and *atole*, which is corn flour, and was beaten until very foamy. The Spanish conquistadors considered the drink to have an unbearable taste: "they (the native people) drink it, but it seems more suitable for pigs than for men." (*History of the New World*, 1564. By **Girolamo Benzoni** [McGee, 2007]).

This concoction was given the name *xoko alth*, which is, translated from Nahuatl, food of the gods. Even so, **Joan Coromines** considers this term to be an erroneous fusion made by the Spanish conquistadors between the Nahuatl terms *pócotl* (kapok seed, also used in the mixture) and *kakawátl* (cocoa drink).

The first European chocolate factory was established in Spain in 1580. It seems that around 1600 some nuns who had settled in Oaxaca (Mexico) began to prepare the drink with sugar and cinnamon, without the pepper and with less foam. This new composition was brought in through the port of Barcelona and the Monasterio de Piedra (the Stone Monastery) in Aragón. From this point it spread to the Spanish court and, in the 17th century, to the French and English ones, with the current name of *chocolate*. The product engendered two variants, both as liquid: the *chocolate a la española*, made with water, and the *chocolate a la francesa*, made with milk. Because of the high price of cocoa, it was a product reserved for society's upper classes, who drank chocolate with all kinds of pastry products at their social gatherings. At the time, drinking chocolate was considered more elegant than drinking tea or coffee. And, moreover, it had a reputation for being a powerful aphrodisiac...

What do dictionaries have to say about chocolate?

In Spanish, the first written reference to *chocolate* is found in Mexico in 1581, and in English in 1604. The term was initially used to refer to liquid chocolate, but soon a dual definition appeared, as a paste and as a liquid, which has remained until today. In some languages, such as Catalan and Italian, which distinguish the gender of words, they give the solid a masculine gender (*el xocolate, il cioccolato*), and use the feminine gender for the liquid (*la xocolata, la cioccolatta*).

How many types of chocolate are there?

If there is a food that covers a thousand variations under the same name, it would be chocolate.

In a general definition, and like all definitions only partially valid, *chocolate is the product obtained by the mixture of sugar and two cocoa derivatives: cocoa paste and cocoa butter*. Here we find the first decisive concept: Chocolate is a substance produced from other substances; it is not obtained as it is from nature but rather is a mixture. The definition does not say what the proportions of its ingredients should be. This means that, in colloquial terms, the name chocolate extends to a very wide family of prepared products, and which have quite varied compositions, textures, physical states and flavors. What all members of this family have in common is that, to a greater or lesser degree, they are derived from **cocoa**. According to the European Union, it must contain a minimum of 35% derivatives from cocoa. The US Food and Drug Administration requires a minimum of 15% of chocolate liquor (cocoa beans fermented, roasted and ground to produce a semi-solid paste). But there are even products incorrectly called chocolate that have no trace of cocoa nor its derivatives: they are *chocolate substitutes* made from carob flour. They cannot legally be called chocolate; certain online shops advertise it as *chocolate without cocoa* with total impunity.

It was in the 19th century that systematic attempts to make the product in solid form began, although before this, in many places and especially in Spain, solid variants of the cocoa dough with sugar were eaten as fast foods, although they were not much in demand due to their rough texture.

In 1828, **Conrad van Houten** designed an industrial process to separate the fat – cocoa butter – from the cocoa beans. And what was left, once ground, was **defatted cocoa powder**, a product that is still being manufactured and sold. By mixing and heating this cocoa powder in the right proportions, with cocoa butter and a certain amount of very finely ground sugar, it was possible – once the mixture was cooled – to prepare the tablets or pieces of chocolate in solid form. The solid product was marketable without the drawbacks of the liquid, and its sale became popular among the population with less purchasing power. The first registered trademark was *Fry and Sons*, English, in 1847. **Daniel Peter**, a Swiss pastry chef, prepared the *chocolate with milk*, using the powdered milk that his compatriot **Henri Nestlé** had successfully made. Milk chocolate was brought out on the market in 1875. In 1879, the Swiss **Rodolphe Lindt** improved the process to obtain the pieces of chocolate when he invented the *conche machine*, which ground the mixture of cocoa beans and sugar much finer, and is the process still used today. The name came from the machine's shell-like shape.

With these processes, dozens of variants of chocolate have been prepared, with all the proportions of cocoa paste, cocoa butter, sugar and – if applicable – powdered milk. One of these variants is called *couverture chocolate*, which is used to coat cakes, fruits and other foods. *Bonbons* are portions of chocolate with different fillings. *Chocolate nougats* (known as *turrones* in Spain) are chocolate tablets with almonds, hazelnuts, puffed rice or a thousand other ingredients in higher proportions than in chocolate tablets, and they are accepted within the regulations under the denomination of *various nougats*.

White chocolate is not chocolate in the original sense because there is no cocoa paste in its formula but only cocoa butter (at least 20%), milk solids and sugar, as well as other ingredients if needed. It was invented in 1923 by the Spanish pastry chef **Santiago Gil** in Guadalajara and later spread throughout the world.

Concerns about obesity or tooth decay and the problem of type I and II diabetes have led to the design of *sugar-free chocolates*, where the sugar is replaced by various sweeteners, as we will see when we talk about ingredients.

There are also many other preparations in which cocoa or chocolate are not major ingredients but are indeed characteristics of the product. Various cocoa shakes have been sold for many years, of which *Cacaolat* is the paradigm and was the first worldwide product of its kind. It was invented by **Joan Viader** (Barcelona, 1931) and is currently produced in different variations. There are also different brands of *cocoa spreads*, such as *Nutella* (Piedmont, 1964-65), among many other brands. *Nutella* was invented as a development of *gianduja*. This product is a mixture of cocoa and hazelnuts invented in Turin in 1806 to replace part of the cocoa in chocolate, which **Napoleon** had ordered blocked because the cocoa came from British ports with which France and Piedmont were at war.

Breakfast preparations with cocoa to dissolve in milk and containing cocoa are *ColaCao*, developed in 1946 in Barcelona by *Nutrexpa*, and *Nesquik*, developed in the USA in 1948 initially with the name *Nestlé's Quik*, and when it was introduced in Europe in the 1950s took on the current name. There are many other brands and variations.

Handmade hot chocolate is the name given to preparations to drink as *hot chocolate*, which contain varying amounts of cocoa. Some are already liquid while others are solid products, usually with rice flour or corn flour. There are many other products with cocoa such as yogurts, flans, cupcakes, coulant, chocolate-covered dried fruits, all kinds of chocolate-coated cookies, and many more. There are, then, dozens of products in which chocolate is the main or characteristic ingredient.

All chocolate-related products are prepared by a manufacturing company, whether we are talking about the individual ingredients – sugar, cocoa – or the finished product, such as a chocolate tablet or a cocoa cream. This also applies to products sold to the public as well as if they are used as ingredients in pastry shops or bakeries. They have their composition, manufacturing, labeling, type of packaging, distribution, conservation and advertising regulated. In the different countries, food codes define, in more or less detail, all the above points.

Product labels

Valuable information for the consumer can be found on the labels. The information on ingredients and nutritional information is especially useful. Advertising and data about the manufacturer, expiration and preferred consumption dates, among others, are less relevant here.

Composition: the ingredients

An *ingredient* is any substance that the producer of the final product has used in its preparation. According to the definition of chocolate, the three main ingredients are *cocoa paste*, *cocoa butter* and *sugar*. But as there may be more, it is obligatory for labels on all prepared products to specify which ingredients have been used, in order from greater to lesser quantity, and this information must be shown in visible lettering on the outside of the product packaging. There are several exceptions to this rule: Most of the ingredients called *food additives*, many of which are added in very small quantities, are placed on the list of ingredients with no regard to quantity, except if they are components present in large quantities. This would be the case for some mass sweetening additives such as *maltitol*, additive 965 (in Europe E-965), which replaces sugar in products without added sugars, such as certain nougats or sugar-free chocolates. In contrast, other sweeteners, such as steviol glycosides (960, commonly known as *stevia*), which are added in very small amounts, are at the end of the list of ingredients, although in advertising it is usually the one in bigger type. Aromas, which are not additives, should be noted generically but without indicating which ones are present in the mixture. If the flavorings have a certain stimulating effect, such as caffeine or quinine, they indeed must be listed on the label on the list of ingredients. Chocolate is considered a stimulant because of the *theobromine* it contains, but since this is a component of the original cocoa and is not added separately, it is not listed on the label.

The ingredients can have many different characteristics. They can be pure chemicals, such as sugar, chemically called *sucrose* $C_{12}H_{22}O_{11}$, or the majority of food additives, such as the emulsifier *lecithin* (322), found in some chocolate bars or in derivative products such as hazelnut creams or soluble breakfast preparations. Other ingredients may be *commercial substances obtained directly from nature*, with the processing needed to be placed on the market, for example chocolate milk or cocoa butter. Both substances are mixtures, but the ingredients are not indicated on the label because no manufacturer has prepared them, but rather they come as such directly from nature. There are other ingredients, however, that are *mixtures with a common name bought from other manufacturers*, and in this case it is indeed necessary to indicate what the ingredients are, and these must be put in parentheses after their name. This introduces complications when it comes to reading labels.

Most of the ingredients used by manufacturers are present in the final product, but there are some cases when they are not. For example, acidity or basicity correctors used to neutralize mixtures, such as *sodium bicarbonate* (500(ii), $NaHCO_3$). These substances react with unwanted components and in the chemical reaction of neutralization both components disappear, leaving in the mixture the product resulting from the reaction, harmless and with no need to be specified. Another example: *Lactase* is an enzyme used to break down milk sugar — *lactose* — into two other sugars, called *glucose* and *galactose*, which can be consumed safely by people who are lactose intolerant. The label must indicate lactase as an ingredient, but not the decomposition products. In the final product we will not find lactase nor lactose, but we will find glucose and galactose, which it is not necessary to indicate on the label.

There are some foods that, for historical or commercial reasons, do not have to indicate the list of ingredients. This is the case of wines and liqueurs in which it should only be noted if they have sulfites (additives from 220 to 228) as preservatives. Chocolate liquors — actually cocoa-flavored alcoholic beverages — don't have to put the presence of chocolate or cocoa on their label.

It is interesting to analyze the ingredients used in products, information provided to us by the labels on consumer products. The law does not require the manufacturer to indicate the quantity of each ingredient, but only the order in which they are present in the formulation. In Table 1 we can see this information for a set of 29 products related to chocolate. These products have been chosen, all of them respected brands in the business and with full health guarantees, just as others could have been chosen: most of the brands make very similar products. Normally, big brands innovate with their formulations and, soon after, the other brands, or the *own brands*, follow them.

On the list of brands analyzed there are various chocolates with a higher or lower percentage of cocoa, other types of chocolates for special uses, such as topping, sugar-free chocolates, white chocolate or milk chocolate, and cocoa derivatives such as cocoa nibs or cocoa butter. In the following section, the meaning of these percentages will be discussed, which cannot be deduced directly from the information of the ingredients. Also included on the list are several products in which the presence of cocoa or chocolate is the most significant element, such as creams, shakes or breakfast preparations, although chocolate is not the main component.

We see how in most chocolates the main ingredient is cocoa, followed by sugar or cocoa butter, in accordance with the basic definition of chocolate. Other special chocolates have milk, sweeteners or other ingredients as the main ones. Conversely, sugar or milk are the main ingredients in derived products (from number 19 onwards).

	PRODUCT	Sugar	Cocoa	Cocoa butter	Milk	Sweeteners	Others
1	Cocoa		1	2			
2	Cocoa nibs		1				
3	Pure cocoa 1		1				
4	Pure cocoa 2		1				
5	Cocoa butter			1			
6	Dark chocolate 99	3	1	2			
7	Dark chocolate 90	3	1	2			
8	Dark chocolate 85	3	1	2			
9	Dark chocolate 70	2	1	3			
10	Dark chocolate 52	2	1	3			4
11	Fondant desserts	1	2	3			
12	Handmade hot chocolate	1	2	4			3
13	Stone ground chocolate	1	2				3
14	Dark sugar-free		1	3	4	2	
15	Sugar-free 47% minimum		2	3		1	4
16	Milk chocolate	1	4	2	3		
17	White chocolate	1		2	4		3
18	Chocolate nougat with rice	1	3	4			2
19	Chocolate 70% with salt	2	1	3			4
20	Classic chocolate shake	2	3		1		
21	Chocolate shake 0.0		2		1	3	
22	Classic chocolate cream	1	3	4			2
23	Original chocolate cream	1	4		3		2
24	Powdered breakfast preparation	1	2				3
25	Intense powdered preparation	2	1				
26	Chocolate dairy dessert	2	3		1		4
27	Classic prepared breakfast	1	2				3
28	Turbo prepared breakfast	1	2				
29	Prepared hot chocolate	2	4		1		3

Table 1. Ingredients of prepared products derived from cocoa, in order of quantity. 1 is the main one, according to the labels. Source: prepared by author.

Composition: nutritional information

Nutritional information must also be included on all labels of prepared products, in accordance with national regulations in each country. While nutritional aspects are not the most essential part of this book, analysis of nutritional information gives us details about composition not only qualitatively, such as the list of ingredients, but also quantitatively, that is, the quantities that are present. Both these pieces of information will help in understanding the structure of chocolate and the other products, and will allow us to enter into aspects having to do with the texture.

Regulations on nutritional information have changed over the years, especially due to pressure from consumer advocates and recommendations from dietitians-nutritionists. This information differs from country to country and is unified throughout the European Union. Nutritional information must have the following sections, referring to 100 g of product:

- Energy value, expressed in kilojoules (kJ) and kilocalories (kcal). The equivalence is 1 kJ = 0.239 kcal
- total fats, distinguishing what amount is saturated fats
- carbohydrates, distinguishing what part is sugars
- dietary fiber
- proteins
- salt

The regulations in Anglo-Saxon countries are somewhat different. The nutritional data is located on a label that is quite differentiated from the rest of the information on the packaging. On this label the nutritional values refer to a portion or serving of the product, specified on the label. The information it contains is as follows:

- energy value in calories, per serving and per container
- total fats
- saturated fats and trans fats
- cholesterol
- sodium
- total carbohydrates
  - dietary fiber
  - total sugars and added sugars
- proteins

Also usually indicated on both types of labels is the percentage of each of the values mentioned in the *population reference intake* (PRI) for an average adult, who is given a daily energy consumption of 8,400 kJ, equivalent to 2,000 kcal. Finally, if the product's advertising highlights that it contains a specific nutrient, such as phosphorus or iron, the amount present should be shown on the list of nutritional information.

Table 2 shows the nutritional information of the 27 products mentioned in Table 1. It is important to emphasize that nutritional information cannot be directly related to the ingredient list, because each of the nutritional components can come from more than one ingredient. For example, fat can come from cocoa butter and from cocoa powder, which contains cocoa butter that has not been completely extracted from it. The relationship is direct only in some cases, like in product 4 (cocoa butter), which is only fat and has no other component.

The NUTRI-SCORE and chocolate

Nutri-Score is a system of classification and evaluation of packaged products that summarizes in a letter and color the nutritional characteristics of a food. It is posted on a label on the front of the package. Each food is assigned A points, the *bad* ones, according to the amount of energy in the food and the percentages of sugars, saturated fats and sodium, extracted from a table of values defined by the system. And it is given C points, the *good* ones, according to the percentages of fruit and vegetables, fibers and proteins. Using the calculation algorithm, the difference between points A and C allows food to be classified into five categories: dark green (A), which have the best nutritional quality; light green (B), yellow (C), orange (D) and red (E), the latter being the ones with the worst nutritional quality.

Nutri-Score labeling is not mandatory and is used only in France, Belgium, Germany, Luxembourg, the Netherlands, Spain and Switzerland, and their derivatives are classified by *Nutri-Score* as having low nutritional quality, which is not good advertising for these products.

The *Nutri-Score* system has been criticized, for various reasons, by many producers and nutritional experts. It is too simplified a procedure, which may be suitable for comparing products within the same category but does not work well for comparing products in different categories. This index does not take into account the presence of food additives or vitamins in the food. For example, a cola soft drink is classified as E, the same drink in its light version as B – less calories – and olive oil as D, but it would be absurd to nutritionally prefer a light cola soft drink to olive oil. This is why the system is continually under review, and many companies are very reluctant to implement it.

	PRODUCT	kJ	kcal	Total fat	Saturated fat	Total carbohydrates	of which are sugars	proteins	salt g	fiber g
1	Cocoa	2534	614	54	34	11	0,5	13	0,05	
2	Cocoa nibs	2480	585	52	30	6	1	12	0,01	23
3	Pure cocoa 1	1360	327	11	7	46	1	25	0,12	32
4	Pure cocoa 2	1412	339	11	6,9	17	1	25	0,05	34
5	Cocoa butter	3766	900	100	60,2	0	0	0	0	0
6	Dark chocolate 99	2365	573	49	30	10	1	14	0,7	
7	Dark chocolate 90	2483	592	55	30	14	7	10	0,03	
8	Dark chocolate 85	2418	584	46	27	22	15	12,5	0,02	
9	Dark chocolate 70	2350	566	41	24	34	30	9,5	0,1	
10	Dark chocolate 52	2227	534	32	19	51	46	6,5	0,1	
11	Fondant desserts	2285	548	34,1	20,2	50,7	46,4	6	<0,01	7,3
12	Handmade hot chocolate	2070	495	24	15	61	50	5,3	0,06	
13	Stone ground chocolate	2117	505	21	12	72	55	5,2	0,01	
14	Dark sugar-free	2024	482	38	23	41	1	7	0,05	
15	Sugar-free 47% minimum	1934	468	33,4	22	51,4	0,6	5,8	0	7,1
16	Milk chocolate	2276	545	32	19	56	55	7	0,24	
17	White chocolate	2652	639	49	37	43	43	5,3	0,21	
18	Chocolate nougat with rice	2179	521	28	10	58	50	7,6	0,18	3,8
19	Chocolate 70% with salt	2348	566	42	26	35	30	7	0,7	10
20	Classic chocolate shake	280	67	1,5	1	10,7	10,6	2,7	0,12	
21	Chocolate shake 0.0	188	45	1,5	1	4,1	3,1	3,4	0,2	
22	Classic chocolate cream	2279	546	32	5,9	58	56	5	0,02	3
23	Original chocolate cream	2252	539	30,9	10,6	57,5	56,3	6,3	0,1	
24	Powdered breakfast preparation	1633	386	3,8	1,6	79,2	75,7	5,1	0,37	7,4
25	Intense powdered preparation	1535	367	11,9	6,8	39,7	29,5	14,6	0,23	21,4
26	Chocolate dairy dessert	515	123	3,3	2	19,5	17,4	2	0,16	
27	Classic prepared breakfast	1593	377	2,5	1,6	78	70	6,6	0,1	7,8
28	Turbo prepared breakfast	1647	389	3	1,5	82	75	5,5	0,3	6
29	Prepared hot chocolate	472	112	1,6	1	22	18	2,5	0,13	

Table 2. Nutritional information of prepared products derived from cocoa, as reported on the labels. Source: prepared by author.

The ingredients

Cocoa

The term **cocoa** refers to three different concepts: the tree, the seeds (cocoa beans) from the tree and the dry powder from these seeds. The tree is the *cacao tree*. In botanical nomenclature it is *Theobroma cacao*, which derives from the Greek *Theobroma*, which means food of the gods. It was given this name by **Linnaeus** himself, inventor of the zoological and botanical nomenclature. The term **cacao** seems to be derived from Nahuatl, from the Aztec language family, one of Mexico's official languages that is spoken by more than two million people. The tree is typical of tropical America, where it is grown widely, and also in Africa and Asia. It needs humid, warm climates and not very high altitudes. The cacao tree produces white flowers that have berries. When the berries are ripe they weigh half a kilo, are about 8 inches in length and 4.5 in diameter and red or yellow in color. The outer shell is 1 or 2 inches, and inside there is an edible pulp that the seeds – between 20 and 50 in each fruit – are immersed in.

These seeds or *cocoa beans* are the basis of chocolate. They are oval, flattened, purple or brownish in color and between 2 and 3 cm in length and 1 to 1.5 cm in width. Their taste when raw is very bitter and astringent. These seeds are wrapped in a white skin. Like all nuts, they have a good amount of fats, protein, dietary fiber and a certain proportion of sugars.

The traditional harvesting process has several stages. The seeds are extracted from the berry with the pulp, are wrapped in banana leaves and left outdoors. The pulp ferments, as well as the inside of the bean, from action by the bacteria present in the environment. The fermentation lasts between three and seven days, and as such it happens that the bean is not germinated. In this phase the temperature increases, the color of the bean changes, and the acetic acid produced by the fermentation inflates the bean, reduces its astringent flavor and its bitterness, and produces the typical cocoa flavors. Then they are dried outdoors, in the sun or with heaters. The dried beans obtained in the end are the product that is sold to chocolate manufacturers.

The completely ground beans are the *cocoa paste*. *Cocoa paste* is usually one of the main ingredients in chocolates. It contains the plant's fats, proteins and carbohydrates, once the beans are roasted and ground. It has a pasty consistency because of the high proportion of fat it contains. When the roasted beans are ground into fragments of a few millimeters, the *cocoa nibs* or *tips* are obtained, which have the same composition as the cocoa beans and cocoa paste. These are products 1 and 2 in Tables 1 and 2.

In the same way as olives are processed to extract their oil, dried cocoa seeds can be pressed and the *cocoa butter* is extracted in the process, product 5 in the tables.

Defatted cocoa is the cocoa bean, ground or not, from which the cocoa butter has been separated. As extraction is usually mechanical, there is always a part of the fat in the dough, typically 11%. Defatted cocoa has the composition of cocoa beans without the fat, that is, carbohydrates (15%), protein (25%) and dietary fiber (more than 30%), in addition to this 11% cocoa butter. This product is crushed and ground until it is an imperceptible powder; most of the particles are less than 2 micrometers in diameter (μm or thousandths of a millimeter), and it is used in the development of different chocolates; it is also marketed under the name of *pure cocoa* or *cocoa powder* for confectionery or for direct consumption. It is listed in Tables 1 and 2 as products 3 and 4.

Fats and triglycerides

Most natural fats, both animal and vegetable, are mixtures of similar molecules, called *triglycerides*, which in a more formal classification should be called *triacylglycerols*. These molecules can be imagined as derivatives of glycerin or glycerol $CH_2OH\text{-}CHOH\text{-}CH_2OH$ in which each OH group has been bonded with a fatty acid molecule, having a generic formula of $COOH\text{-}(CH_2)_n\text{-}CH_3$. Fatty acids have chains of 12 to 18 carbon atoms. The chain can be *saturated*, like the one in the previous formula, without double bonds between carbons, or it can be *unsaturated*, with a double bond inside, as in the case of *oleic acid* $C_{18}H_{34}O_2$, which has the following structural formula: COOH-$(CH_2)_7$-CH=CH-$(CH_2)_7$-CH_3.

Each triglyceride molecule can derive from three different fatty acids, and this variety produces the plurality of existing triglycerides, which is enormous. Figure 2 shows an example of triglyceride.

Cocoa butter

Cocoa butter is nothing more than 100% fat. It is product 5 in Tables 1 and 2. It is a mixture of fatty molecules called *triglycerides*, of which 60% are saturated fats. The product has almost no taste or odor, a yellowish color, and is sold to make white chocolate as well as to add to dark chocolates in the proportions necessary for its preparation. The most common triglycerides in cocoa butter derive from palmitic acid (abbreviated as P, with formula CH_3-$(CH_2)_{14}$-COOH), stearic acid (S, CH_3-$(CH_2)_{16}$ -COOH) and oleic acid (O, CH_3-$(CH_2)_5$-CH=CH-$(CH_2)_5$-COOH). Any of these acids may have bonded to any of the –OH terminals of the glycerol, but some combinations are more frequent than others. Thus, it has been observed that between 85 and 90% of cocoa butter consists of only three sets of triglycerides: POS (palmitic, oleic and stearic), SOS and POP. These specific compositions are relevant because they will define the interval of melting points of the cocoa butter, as we will see when talking about texture.

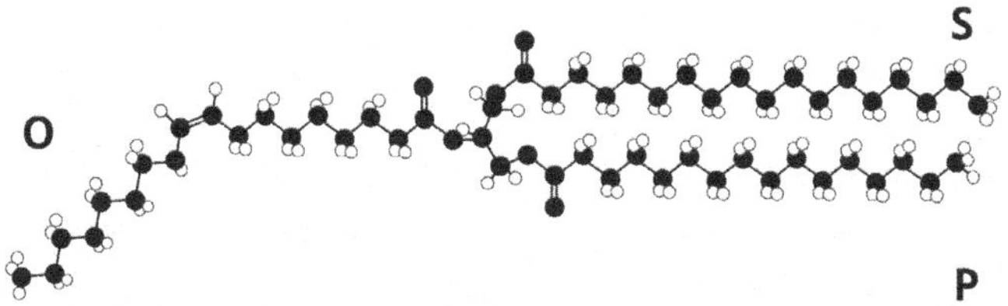

Figure 2. POS triglyceride molecule. Three branches come from the central glycerol molecule. The two on the right are stearic acid (S) and palmitic acid (P) chains, both saturated. The chain on the left is unsaturated oleic acid (O), as shown by its double bond in the center of the chain. Source: prepared by author.

Sugar

The term *sugar* applies, in the chemical context, to a broad set of compounds in the carbohydrate family. In the food context, however, the term *sugar* refers specifically to *sucrose*, with the formula $C_{12}H_{22}Or_{11}$. It is present as this molecule in sugar cane and beets, from where it is obtained by grinding the vegetables and extracting the sugar with water, which makes it crystallize. Refined sugar is generally used, and is called *white sugar*. It is probably the purest product we have in our home. But unrefined *brown sugar* can also be used in some mixtures. For its use in chocolate and other products, it must be ground to obtain a fine powder, known *as icing sugar or glaze*, which contains 2% to 3% starch. Its particles are usually micrometrical in size, just like cocoa powder.

Milk

Powdered milk, either whole or skimmed, is used to make milk chocolate; depending on the final mixture desired, the resulting flavor and texture will be different. Dairy products with chocolate, such as shakes, use liquid milk, whey, and other similar derivatives.

Fats not derived from cocoa

In the formulation of products such as cocoa creams (products 21 and 22 on the tables) and others, it is authorized to use non-cocoa vegetable fats, such as *coconut oil* or *palm kernel oil*, the latter extracted from the almond of the Guinea oil palm. These are oils with a high proportion of saturated fats. The latter is also used as an additive in cocoa butter to check the phenomenon known as fat bloom, which will be discussed in Chapter 3. *Shea butter*, with more than 60% of its triglycerides unsaturated, derivatives from oleic acid, is also used. *Rapeseed oil* and *palm oil*, which most producers now prefer to avoid because of the poor image they have among certain consumers – an image, by the way, that is totally unjustified – have also been used. These fats are also used in certain chocolate preparations.

Emulsifiers and thickeners

Lecithin, or rather lecithins, is a family of molecules of the *phospholipids* type, that is, fats with phosphorous atoms in their composition. The most typical is *phosphatidylcholine*. It is found in egg yolk and soy beans, where it is primarily obtained. The lecithins family is the food additive E-322, widely used as a stabilizer and emulsifier in various prepared products. In some chocolate formulations lecithin is added to homogenize the mixture, as will be discussed later.

Also used as thickeners and emulsifiers are E-407 *carrageenans* and the *mono-* and *diglycerides* of fatty acids E-471. All these food additives are explicitly listed on the ingredients list, with the name or number.

Sweeteners

Concerns about tooth decay, obesity and the growing prevalence of type II diabetes in the population, especially among young people, have favored the tendency to replace sucrose – sugar – with other sweeteners that do not have these health problems. These are food additives with codes E-950 to E-970, in addition to E-420 (sorbitol) and E-421 (mannitol), which are not present in chocolate.

Bulk sweeteners replace sugar. They are added in certain amounts and add weight, texture and volume to the product, as well as sweetening it. In chocolates without added sugars, the sweet taste usually comes from adding *maltitol.* It is sweetener E-965 (i), with the chemical formula $C_{12}H_{24}O_{11}$. Chemically it is a polyalcohol, obtained from the *maltose* derived from starch. It has a sweetening power of 75 to 90% of cane sugar, does not cause cavities, has half the calories of sucrose, and in the body it is not transformed into glucose so it can be used by diabetics. In appearance it is a thick white powder that gives body to preparations where it substitutes sugar, such as in chocolates, gum and nougat.

Sometimes sweeteners other than sugar and maltitol, such as *erythritol*, E-968, are also used. Some formulations use *steviol glycosides* (E-960) extracted from stevia, which always accompany maltitol because these glycosides are not bulk sweeteners and would not provide thickness to the chocolate. This substance, with high sweetening power, has become fashionable in recent years, even though there are no convincing reasons to prefer it, beyond a certain halo of being ecological.

Other ingredients

Stone ground chocolate (or *chocolate with flour*) is a classic solid preparation used to make *handmade hot chocolate*, a paste-like liquid preparation that must be cooked before being consumed. It contains a flour as an ingredient that is usually *rice flour* or also *corn starch*, to thicken the mixture.

Other ingredients can also be added to give flavors to the chocolate. *Cinnamon*, *vanilla* or *vanillin* ($C_8H_8Or_3$) are generally used. The latter is a compound of the vanilla pod and can be obtained by chemical synthesis. Other vanillin derivatives can also be used.

Sodium bicarbonate (500(ii)) is used in certain mixtures to neutralize its acidity.

Some chocolates add *salt* (NaCl sodium chloride) to give it a different flavor (see product 18 in Tables 1 and 2).

The imagination of producers in search of new markets and consumer segments means that all kinds of chocolate tablets are made with nuts, other fruits, all types of liqueurs, pepper and a long list, even longer if we add the highly creative chocolate nougats, indistinguishable from chocolate tablets but with their own section in the food codes. One of the oldest examples was product no. 17 in Tables 1 and 2, with puffed rice.

Chapter 2

The composition. Chemical analysis and microscopy

Throughout the previous chapter the review of chocolate products has only taken into account the qualitative and quantitative inventory of the ingredients used to make the mixtures, which is mandatory on the packaging. Chemical analyses will let us know the products in more detail and give us more in-depth information.

There are several types of chemical analyses. *Elemental analysis* of chocolate formulations is the determination of the ratio by mass of the *chemical elements* that make up the different ingredients: *carbon* C, *oxygen* O and *hydrogen* H in fats and carbohydrates, also *nitrogen* N in proteins, *sodium* Na from salt, and other lesser elements such as *phosphorus* P or *magnesium* Mg. This analysis is of direct interest in the field of nutrition and dietetics but does not give us information on the organoleptic properties of the substances. Elemental analysis is not included on product labels.

Molecular analysis is the determination of which different molecules are in the formulation and how many of each there are. These molecules are the ones that were already in the ingredients because when we make the mixtures they are not created or destroyed: the preparation of chocolate and of other preparations with chocolate is not done with chemical reactions, but rather these are physical processes of a change of size (grinding), change of state (melting or solidification) or mixtures, more or less deep, such as dissolutions or emulsifications. There have indeed been chemical reactions when fermenting and roasting the cocoa beans, but the molecules produced in these processes are the ones that will be present in the cocoa paste.

Most of the ingredients in chocolate formulations are derived directly from natural products and are, as already mentioned, very complex mixtures. In fact, cocoa *fats* are natural mixtures of hundreds of different, but very similar, triglyceride molecules; the *proteins* present in chocolate are also complex molecules derived from the twenty existing essential amino acids; the natural *carbohydrates* present are *sugars* and *dietary fiber* present in cocoa paste, of various molecular compositions; the added *sugars* are usually *sucrose*. Finally, there are all the additives, which are mostly pure molecules. The molecules in the mixtures are responsible for three important organoleptic properties: the smell, the taste and, partially, the color. Conversely, knowledge about molecules does not give us information on the texture of the compositions, and it is necessary to know their microscopic structure.

In the composition there are also a large number of molecules in very small quantities, all derived from natural products. These products that are present in small quantities or *micronutrients* are of many types, and as research progresses more and more are discovered. Standing out, for example, are various flavonoids, tryptophan, and derivatives from magnesium, potassium, cadmium or lead. This will be discussed in Chapter 4. There are other types of chemical analyses that are usually done to characterize foods, but these will not be discussed here.

Chocolate under the microscope

A pure substance such as water or sucrose, seen under an optical or electron microscope, does not allow differentiated particles to be distinguished: its molecules are too small to be seen. Same for solutions of a salt or sugar, because they dissolve completely, and the molecules of sugar or salt ions remain among the water molecules and are so small and so separate that the microscope does not have enough resolution to distinguish them. Instead, there are more or less complex mixtures that have some of their components grouped in the form of droplets or clumps, which are made up of millions of molecules and are not dissolved in the overall mass but rather dispersed in the mass, but each droplet maintains its identity. These are what are known as *dispersed systems* or *colloidal systems*, which are of decisive importance in biology, chemistry and food science.

An example of these systems is mayonnaise, which is made up of oil droplets — each droplet with millions of triglyceride molecules — dispersed in a continuous medium which is the water in the egg white. Butter is another example of a dispersed system, where there are droplets of water — each with millions of water molecules — dispersed in the mass of fat, which are the triglycerides from the milk. Mayonnaise and butter are examples of *emulsions*, the first of *oil in water* (in its technical abbreviation, O/W) and the second of water in oil (W/O). The terms "water" and "oil" are generic, to indicate the aqueous phase, with a predominance of water, and the oily phase, with a predominance of oil: they do not refer to pure water or oil. Let's repeat that *pure oil* from the producer's point of view refers to it not having undergone adulteration or additions. But *pure oil* is not a pure substance from the chemical perspective but rather a natural mixture of many different triglycerides, and each one is indeed a pure substance. If one of the phases is solid, we talk about suspensions: For example, bechamel sauce is a suspension of solid particles of flour – pieces of a grain of cereal – in the milk-butter mixture.

Chocolate is also a rather complicated dispersed system, as seen in Figure 3. We find the three essential components: sugar, cocoa powder and cocoa butter. Solid sugar particles are crystalline, finely ground, and separated from each other. They are crystals formed by sucrose molecules. Cocoa is made up of pieces of ground cocoa beans, in the form of a very fine powder, and the particles are separated from each other. Each of the cocoa powder particles consists of dietary fiber, some sugars, proteins and a little cocoa butter, just as they were in the original cocoa bean cells. And finally, cocoa butter is made up of mixtures of triglyceride molecules, and is solid or liquid depending on the temperature at which it is and according to its final composition (see Chapter 3). As a large part of the cocoa butter mass is melted, these are not particles but rather is a very viscous liquid and constitutes the continuous phase of the mixture.

How are these three components structured in chocolate? Sugar particles are sucrose molecules. These molecules would dissolve very well in water because they have many atomic chains with --OH endings, which are very *hydrophilic*, that is, they have an affinity for water. But there is no water in chocolate. Instead, cocoa particles, from which much of the fat has been removed, are *hydrophobic*, they have a very dry surface, like flour, and it would be difficult for them to get wet. The cocoa butter present in the mixture, which is a fat, easily disperses the cocoa particles and, conversely, has no affinity for sugar. In this sense, the process of preparing chocolate has a certain similarity with the preparation of bechamel sauce: in this case butter – the fat – is heated and flour is slowly added, which is dispersed and roasted a bit. The result is a golden paste that the French call *roux*, which dispersed with milk will make the sauce. Similarly, in chocolate there is a deep mixture of cocoa powder and cocoa butter, with sugar crystals scattered throughout the mass.

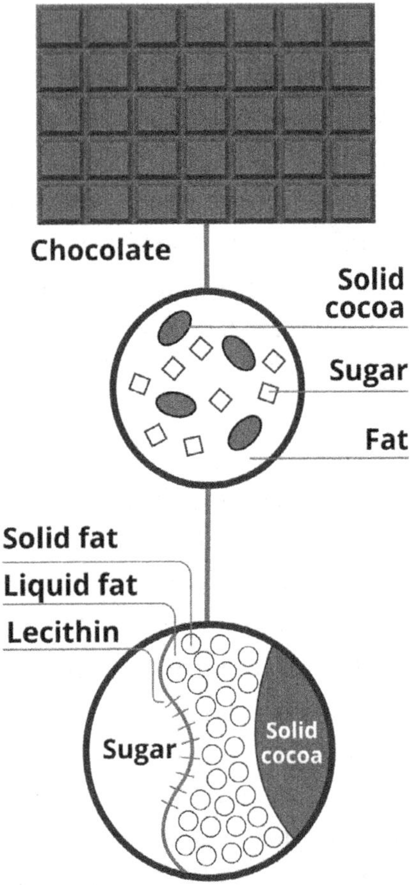

Figure 3. Structure of chocolate seen with successive microscopic enlargements. The chocolate tablet looks homogeneous, but a first enlargement lets the sugar crystals and cocoa powder particles immersed in the cocoa butter mass be seen. A greater enlargement would perhaps allow us to see that the cocoa butter is partially melted and there are solidified butter crystals. If it contained lecithin, its molecules, invisible under the microscope, would be partly attached to the cocoa butter mass and partly to the sugar, favoring its dispersion. Source: prepared by author.

In order to make the solid dispersion of cocoa powder in cocoa butter accept more readily the added sugar – hydrophilic – sometimes a little lecithin, which is additive 322, is added to the mass. Lecithin molecules have two tails, one hydrophilic and the other hydrophobic. The hydrophilic end adheres to the sugar particles, and the hydrophobic end to the fat, favoring the mixture's dispersion and the result is a stabilized solid suspension.

As has been said, depending on the temperature, cocoa butter will be partially solidified because, as mentioned, cocoa butter is a mixture of many triglyceride structures with different melting points. Therefore, the cocoa butter matrix is liquid with solid cocoa butter crystals dispersed inside. Figure 3 represents the overall dispersed system. Microscopy, however, does not let us distinguish liquid cocoa butter from solid cocoa butter, and other techniques must be used such as magnetic resonance imaging (MRI), well known in medicine, which allows soft masses – liquid cocoa butter – to be distinguished from hard masses, the solid crystallized cocoa butter.

This entire structure will largely determine the texture of the final product.

Some physical properties of chocolate

Density

Many people would say that handmade hot chocolate is very dense. But they would be speaking in colloquial language. They could also say it is a thick, pasty or viscous substance. But, in technical and scientific usage, **density** is the mass – the kilos – of substance that is in one cubic meter of it and has nothing to do with whether the substance is fluid or pasty. These other properties have to do with the *viscosity*, and will be discussed later. There are substances that have low density and are very viscous, such as oils, and other substances that are very dense and not very viscous.

Density is not an additive property. That is, the density of chocolate cannot be known by a simple calculation based on the density of its components because it depends on how the mixture is structured. The density of cocoa butter is approximately 850 kg/m^3, much less than the density of water, which is 1000 at room temperature. Sugar is the densest component of chocolate, with a value of 1587. The density of cocoa powder is between 400 and 600, but this is its apparent density as it is not compacted, with all the air included that exists between the particles. The true density of the solid cocoa particles is above 1000 kg/m^3. The three ingredients mixed in the chocolate will produce a product having a density that will depend on the proportions of each component: Generally speaking, the darker the chocolate, the less sugar there is and, therefore, it has less density. Chocolates have densities between 1200 and 1300 kg/m^3. This latter density refers to milk chocolate, which usually contains a higher proportion of sugar. All chocolates are denser than water: they do not float.

Electrical conductivity

When talking about the microscopic structure of chocolate it has been seen that it is a solid dispersion, in which the dispersed phases are the sugar crystals and cocoa powder, and the continuous phase is the cocoa butter, which does not conduct electricity. Thus, chocolate does not conduct electricity.

Hardness

The **hardness** of a substance is a property that is measured on what is known as the **Mohs Hardness Scale**. This scale, created in 1825 and one of many that exist, consists of a set of ten reference minerals of increasing hardness that are used to test a material's resistance to being scratched. The softest material on the scale is talc, a mineral given a hardness of 1, and the hardest is the diamond, with a hardness of 10. The surface of chocolate is mostly cocoa butter, which has very low hardness, like all solidified fats: it can be scratched with a fingernail at room temperature. Therefore, it has a hardness of less than 1.

Fracturable

Fracturable is the property of a material to break easily, and is completely independent of hardness. Fragility depends on the structure of the material. Chocolate has the cocoa butter as a continuous phase, and at room temperature this mass can be cut or broken easily because it does not deform: it is quite fragile and makes a little cracking noise when it breaks. By contrast, chocolate tablets that are refrigerated – a practice not recommended as the properties of the cocoa butter change, as we will see – are somewhat harder and it is a bit more difficult to break them. In both cases they break clean, but the break is not regular like how a crystallized mineral would break.

Wettability

Wettability is the capacity of a material to have water stick to its surface. As the surface of chocolate is oily, it does not get wet easily: it is a *hydrophobic* surface, as was mentioned when talking about the microscopic structure. The water that can reach it is in the form of large drops that can be easily removed.

Viscosity

Viscosity is the measure of the difficulty a fluid has to flow. This concept is part of a scientific and technical field, **rheology**, that has huge importance in food technology. In the world of chocolate, studies on viscosity are important in liquid or pasty derivatives. Its relationship with texture will be discussed in Chapter 3.

Melting point

Chocolate cannot be given a specific melting point. This question is very important when talking about the texture of chocolate products, and a later section will be devoted to it.

Classic organoleptic characteristics such as color, smell and taste will be described and discussed in Chapter 4.

Chapter 3

The textures of chocolate

Texturization

In the world of gastronomy, one of the most successful concepts used by innovative restaurateurs is *texturization*. It does not mean, unlike what it may seem, giving texture to a product that does not have it, since everything has a certain texture. Rather, it means modifying the texture that it currently has and giving it another one considered to be more desirable. The first problem in this question is the definition of texture. And once the texture of the product is defined and characterized, we must then invent the procedures to modify it and get the desired new texture.

What texture does chocolate have? This question asked of the general public can have many different answers, such as smooth, delicious, sweet, bitter, crunchy, dark, etc. These open and spontaneous answers help us approach the subject of textures scientifically since they are the basis of *sensory analysis*, which will be discussed later. The concept of *texture* has changed a lot over time, as can be seen by consulting historical dictionaries, and has continually expanded. It began by referring to the feel of clothes and fabrics on the skin, with attributes such as roughness, coarseness or warmth. Later all the characteristics related to the feel of any material were included, such as a mineral, plastic or metal, and attributes such as graininess, degree of polish or hardness were added. Currently, the concept of texture is also linked to food and most especially to its gastronomic aspects, with attributes such as crunchiness, creaminess or oiliness. By extension, we also talk about the texture of a piece of music, the texture of the floor, or the texture of a painting, among many other accepted meanings. In all fields where the concept is used, however, it never refers to a single property but rather to an entire set. In textures, the sense of *touch* is fundamental, and that is why the study of textures is part of *haptic*, a now standard neologism that means the *science of touch*.

The definition of texture has different meanings, usually referring to woven and non-woven fabrics. Over time the concept of texture in reference to food has been introduced, and is now widely used. In a general sense this refers to the set of mechanical, geometric and other properties of a product, perceptible by the mechanoreceptors, the tactile receptors, and in certain cases the visual and auditory ones. Texture is, therefore, a set of properties — not just one — that is perceptible by different sensory receptors in the human body.

In order to advance in a deep understanding of why chocolate has the properties it has, it will be necessary to relate the physical and chemical characteristics with the terminology, concepts and ways of measuring the different properties that are linked to the texture, and with the arrangement and characteristics of the particles that make up the chocolate. We have seen the basics in Chapters 1 and 2. Now we must do the integration, and this will be the goal of the present chapter.

Measuring sensations

The quantitative measurement of sensations in tasting a food, *its organoleptic characteristics*, is difficult to objectify and standardize, as sensations are personal and expressed in terms that resist objective evaluations that are comparable between individuals. Let's take an example: Imagine that three consumers evaluate a chocolate with 70% cocoa (like product number 9 on Table 2); the first consumer might say they find it sweet and like it; the second one finds it sweet and doesn't like it; and the third one doesn't think its sweet, but still likes it. The chocolate is the same, with 30% sugar, which is an objective value measured by chemical analysis techniques, and thus as a value is non-debatable. But, conversely, the two properties analyzed on the tasting panel, which are the degree of sweetness – the *organoleptic* characteristic – and the satisfaction provided by the product – the *hedonistic* sensation – are not unanimous.

Thus it is necessary to find procedures to unify the terminology of the non-objective properties – sweetness, in the previous example – related to texture and link them with objective physicochemical observations established in the laboratory. Then, comparative scales must be found for each of these non-objective properties. And finally, we must find systems – preferably quantitative – so consumers can express their assessments of the product's properties and that these evaluations are objectively comparable with each other. This is the task of *sensory analysis*. All this work has different final objectives: quality control of products; knowing the acceptance that a new product – or an old one that has been modified – has in the eyes of consumers; and to link all the prior work with scientific and technical measures that allow researchers and producers to define the most objective procedures to compare products.

The terminology of textures

Anyone can describe a food with everyday terminology. For example, a tangerine can be said to be acidic, soft, hard, fibrous, juicy, compact and a thousand other adjectives. A chocolate, to be sweet, floury, with pieces of fruit, crunchy, soft, brittle, etc. Systematically collecting all the adjectives that hundreds of consumers say is not easy: All consumers need to refer to the same feeling with the same adjective, and redundancies and inaccuracies must be eliminated. This work has been done in all fields, including in all food sectors, and it has been possible to unify the terminology used to describe textures. The different state agencies create rules on how to make analyses and measurements of all kinds of products, materials, instruments and everything imaginable. And, naturally, there are rules for defining how to do sensory analyses and determine the textures of food products. The international standard 11036:2020 refers to sensory analysis and texture profile and is the one in force at the time of writing this text.

According to this standard, food textures are divided into three large independent groups corresponding to three types of properties: *mechanical properties*, *geometrical properties* and *other properties*. Each of these groups of properties contain various *attributes*, which are each one of the properties; and each of the attributes has several *descriptors*, which are the different adjectives that specify the properties. There are in total up to forty descriptors. For each one of these descriptors, the standard gives an example of a food. The entire diagram is summarized below:

Mechanical properties

- Hardness
  - Soft (spreadable cheese)
  - Firm (olive)
  - Hard (hard candy)
- Cohesiveness
  - Fracturable
    - o Crumbly (shortbread cookie)
    - o Crunchy (apple, raw carrot)
    - o Brittle (roasted peanuts, bread crust, potato chips)
  - Chewiness
    - o Tender (green peas)
    - o Chewable (gummy bear)
    - o Chewy (tough meat)
  - Gumminess
    - o Gritty (fiber cookies)
    - o Floury (cooked dried beans)
    - o Pasty (mashed potatoes)
    - o Gummy (gelatin)
- Viscosity
  - Fluid (water)
  - Thick (instant hot chocolate)
  - Viscous (condensed milk, honey)
- Springiness
  - Plastic (butter)
  - Elastic (squid)
- Adhesiveness
  - Sticky (over-cooked rice)
  - Adhesive (coffee caramel)

Geometrical properties

- Graininess
  - Floury (icing sugar)
  - Grainy (semolina)
  - Gritty (certain pears)
  - Lumpy (cottage cheese)
  - Pearly (caviar)
- Structure
  - Thin layers (cooked cod)
  - Fibrous (celery, asparagus)
  - Pulpy (peach pulp)
  - Cellular (mandarin)
  - Spongy (merengue)
  - Crystalline (granulated sugar)

Other properties

- Moisture
  - Dry (salted cracker)
  - Moist (certain apples)
  - Juicy (orange)
  - Succulent (juicy meat)
  - Aqueous (watermelon)
- Fatty nature
  - Oily (sardines in oil)
  - Fatty (fried bacon)
  - Greasy (tallow)

The same food may have attributes from different properties to characterize it, but only one descriptor for each attribute. This diagram has different weaknesses, such as the repetition of some attributes for different categories, or some examples that are not sufficiently specific ("certain pears," for example). Despite this, it is broad and comprehensive enough to cover most of the food textures. We can also see that instant hot chocolate is mentioned in the mechanical properties, specifically in the viscosity attribute, with the descriptor *thick*.

Descriptors of chocolate textures

The basic ingredients of chocolate each have very different textures. At room temperature, and only by touch but not in the mouth, the descriptors that can be used for them are very different from each other:

- Powdered sugar can be said to be hard, brittle, grainy or floury, crystalline and dry.
- Conversely, cocoa butter is soft, chewy, pasty, viscous, plastic, sticky and greasy.

- Cocoa powder, on the other hand, is crumbly, gritty, floury and dry.

However, as with many other properties when the ingredients of a preparation are mixed, the mixture's final product ends up having textures that do not necessarily resemble those of its ingredients. Perhaps this is the essence of cooking: mastering the art of the textures within a mixture.

Furthermore, the texture a chocolate tablet has when it is taken out of the wrapper is not the same one we enjoy when we put a piece of chocolate in our mouth, it changes while in there, and we finally swallow it. This aspect, which, in fact, is essential in gastronomy, will be described in this chapter a little later on.

We will refer to classic chocolate tablets: only cocoa, cocoa butter and sugar (products 6, 7, 8, 9, 10 and 11 on Tables 1 and 2). The descriptors that can be used with them, at room temperature and only by touch, without ingestion, would be:

Mechanical properties

- Hardness: solid, in general
- Fracturable: crunchy or brittle
- Gumminess and viscosity: these cannot be assessed directly with the naked eye
- Elasticity: plastic
- Adhesiveness: predominantly sticky

Geometric properties

- Graininess: not grainy
- Structure: rather crystalline

Other properties
- Moistness: dry
- Oily nature: not at room temperature

These descriptors of the texture of solid chocolate can be nicely associated with the microscopic structure of chocolate. It is a solid dispersion in which the continuous phase is cocoa butter and the presence of powdered sugar and cocoa powder does not excessively change the overall textures. At temperatures slightly cooler than room temperature, cocoa butter is a solid mass with embedded sugar and cocoa particles. But it is not as simple as the above phrase might make it seem. This is because cocoa butter is a very complex product.

Fusion and solidification of cocoa butter

This is a decisive aspect to understand the properties of chocolates. As we have seen in a previous section, cocoa butter is a mixture of many different triglycerides, mostly triglyceride POS (derived from palmitic, oleic and stearic acid), SOS (stearic, oleic and stearic acid) and POP (palmitic, oleic and palmitic acid). In fact, it is a good bit more complicated, because the base glycerol structure of triglycerides has three carbons with OH groups that each react with a fatty acid. And depending on the order of the fatty acids bound to the glycerol, the properties of the molecule change: triglyceride POS (with the palmitic bound to carbon 1, oleic to carbon 2 and stearic to carbon 3) is not the same as triglyceride PSO: the shapes of the molecules are somewhat different, and the solid structures they form when crystallizing are also different.

Figure 2 is the representation of a typical POS triglyceride, with its three P, O and S chains developed. The triglycerides POS, SOS and POP make up more than 85% of cocoa butter. P and S are saturated chains, that is, all the carbon-carbon bonds are simple bonds. In contrast, the oleic acid chain O has in its center a bond between two carbons that is a double bond, an unsaturated bond, which gives it different properties: a higher proportion of chains with unsaturated bonds causes the fats to have lower melting points It is the same thing that happens in pork fat as compared to lamb fat: the latter is mainly from saturated chains, and that is why when cooled it becomes solid more quickly. In the case of cocoa, the curious fact is that the closer the fruit's origin is to the equator, the softer – less saturated – the cocoa butter it contains. This is important for chocolatiers who make chocolates and bonbons. For commercial chocolates, it is of interest that the oleic acid has mainly bonded to the carbon 2 – the central one – of the glycerol, like in the figure.

Solid or liquid?

The elementary classification of the states of matter in solid, liquid and gas is not appropriate for food, nor for chocolate, for two reasons. Dispersed systems, and chocolate is one of them, consist of different phases, each of which may be in a different state: In chocolate, sugar is in the form of solid crystals, cocoa powder as solid particles as well, and cocoa butter, at room temperature, is a mixture of solid crystals and a liquid mass. The chocolate tablet, which is the mixture of all the previous phases, is seen as being solid because it maintains its shape, but inside there is liquid: it is a complex dispersed system. What is more, cocoa butter, when melted, provides highly viscous liquids that we find difficult to put in the category of liquid as they are very different from other liquids in the kitchen, such as water, vinegar, milk or wine. But cocoa butter or melted chocolate are liquids, like very viscous oils. They are really a bit peculiar: we will see their characteristics when talking about viscosity and viscoelasticity.

Polymorphism

Mixtures of chemically similar substances having somewhat different molecules, such as these triglycerides, usually do not have a single defined melting point, but apparently a range of melting points. This is because these triglycerides show a phenomenon known as *polymorphism*. This term refers to the phenomenon in which, when cooled and frozen, the mixture's molecules group together into different types of structures called *polymorphs*, with packing of varying degrees of compaction but each of them relatively stable. Each polymorph behaves almost like a pure substance: it has specific properties different from the other polymorphs, can crystallize with its own structures, and has a defined melting point that is different from the other polymorphs. This behavior is not specific to cocoa butter; other substances also show it. The simplest and most well-known case is that of graphite, diamond and graphene polymorphs: they all have the same chemical composition, are forms of pure carbon C, but they are not similar at all in their properties. Not all materials that crystallize present the phenomenon of polymorphism, but in cases where it occurs, knowledge of the phenomenon is critical: For example, certain polymorphic forms of drug molecules are toxic and, conversely, others with the same composition but crystallized in a different polymorph are healing drugs. It is necessary, when the drug is manufactured, that the non-toxic is crystallized and crystallization of what is harmful is avoided.

The behavior of the polymorphs from the cocoa butter is also critical for chocolate's properties. Six polymorphs have been identified in solid cocoa butter, which are called, showing very little imagination, polymorphs I, II, III, IV, V and VI, according to a classification done in 1966. Each of these polymorphic forms has its own melting point. They are, respectively, 17°C, 23°C, 25°C, 27°C, 33°C and 36°C. This fact explains why, at a room temperature of 20°C, part of the cocoa butter in the chocolate is in liquid form (polymorph I) and there are lumps on the inside – imperceptible to the palate – of solid cocoa butter, corresponding to the other polymorphs that have not yet melted. Liquid cocoa butter, at the temperatures mentioned here, is a very viscous liquid that does not flow easily due to its structure. At first glance, the liquid is not visible. Crystallographers explain that, starting from liquid cocoa butter – whether pure or already mixed with cocoa and sugar in the form of liquid chocolate – and is thus at a temperature above 36°C, the speed at which we cool the mass allows the different polymorphs to be obtained separately. And this is an essential aspect when it comes to making chocolate tablets or other forms. Moreover, with time and depending on the room temperature, some polymorphs transform into others, an aspect that is important in the storage of cocoa butter and chocolates.

The desired polymorph in chocolate bars and tablets is polymorph V because it has a melting point of 33°C, slightly lower than the temperature inside the mouth cavity, which is 36.5 or 37°C. When you put a piece of chocolate in your mouth, if the cocoa butter is mainly polymorph V, your mouth will heat the chocolate and it will melt, while the palate and tongue will cool slightly, providing the heat required for it to melt. This refreshing sensation is pleasing while eating the chocolate and is what producers look for in their products. In addition, the V form is harder and when chewed is quite crunchy, another attractive aspect.

How to maximize getting polymorph V from melted cocoa butter? Controlling how fast the mass cools is decisive in the process: if the cooling speed is 10°C per minute or faster, mainly polymorphs I and II are formed. If the cooling rate of the mass is 2°C per minute, principally polymorphs III and IV are formed. None of these polymorphs interest the manufacturers. To obtain the maximum amount of polymorph V by cooling, it would be necessary to refrigerate the mass of melted cocoa butter – or liquid chocolate – at speeds as slow as 0.01°C per minute or less, which is industrially unfeasible. However, there is an alternative process called the *tempering process* that lets the polymorph V crystals be obtained and stabilized. The *tempering process of the chocolate* starts from completely melted chocolate, at more than 45°C. It is then cooled quite rapidly to 27°C, just before the polymorph IV crystallizes. Then, it is reheated for 15 minutes at 32°C and is kept at this temperature for a long period of time, called *maturing*. The growth process of polymorph V crystals takes place mainly during these two cooling and maturing processes. This is how the desired texture is achieved. Additionally, the surface of the solidified chocolate will be shiny and smooth, also desirable properties in chocolate. The tempering process can also be done in your kitchen at home, provided that the temperatures are precisely respected.

Fat bloom

Fat bloom is names given to the phenomenon of a layer of white fat that appears on the surface of solid chocolate. The phenomenon occurs mainly in summer when chocolate at a somewhat high room temperature has been subjected to sudden cooling in the refrigerator. Polymorphs with a lower melting point are melted at room temperature and can spread by capillarity through the interior of the solid structure and reach the surface. If the whole is cooled, they freeze on the surface and give it a white appearance. This newly solidified fat usually does so in the form of polymorph VI, which has a rough and unpleasant structure, in the opinion of many consumers. Chocolate in these conditions has the same original nutritional properties and flavor, but many consumers don't like it. There are brands that stop selling some of their products during the summer, to prevent them from showing *fat bloom*, which some people imagine – erroneously – is due to mold. Other brands have designed *antibloom* fats, which are added to the cocoa butter and stop polymorph V from becoming polymorph VI.

Texturometers and sensory analysis

When we melt chocolate, different physical and chemical phenomena occur that can be interpreted based on the properties we have seen on solid chocolate and its melting, but it is necessary to go a little deeper. Specifically, it is time to talk about properties such as *viscosity* and *viscoelasticity.* It is also necessary to talk about the devices that measure them, such as *viscometers* and *rheometers.* And, above all, it is necessary to know how textures are measured, and for this it is also necessary to talk about *texturometers* or texture analyzers and other additional equipment. And, after all this, we will be able to understand what happens when making melted chocolate, and why it almost always burns when we make it at home.

As seen in the previous section, there are many components when it comes to evaluating a texture. The lists of attributes are not quantitative but rather qualitative, and it is interesting for producers and researchers to have systems that are as objective as possible to determine the components of textures. To this end, there are *texturometers*, which, as the name suggests, are devices to measure and quantify textures. Viscometers and rheometers are texturometers specialized in measuring the properties of fluids.

These devices have been used for many years in the fields of materials science and technology under the name of *universal testing machines.* They quantify how a food behaves when it is subjected to a mechanical disturbance. Its application in the field of food became popular in the 1990s. Their design is adequate to obtain technical data on the behavior of materials, but it is not easy to relate the data they provide with the most basic scientific physical properties and with the sensations of the consumer.

A texturometer is a table-top instrument that is essentially made up of two parts. The first one is fixed, and it is where the material you want to study is deposited. It has different sensors to determine material properties such as dimensions, temperature, degree of deformation, the pressure to which the sample is subjected and others. The second part of the device is a moving arm that is positioned over the material and submits it to a mechanical action, with one of the heads that can be fixed on it. This action performed by the head, depending on the parameter to be measured, can be pressure, a cut, rotation, shaving, deformation, a perforation, puncture or any other type. The apparatus as a whole simultaneously determines and records the force required to perform the action and what the response obtained has been, such as the degree of penetration, the deformation obtained or the crumbling achieved, to mention three of the many aspects that can be measured with the appropriate accessories. Analysis of the results lets the food be characterized and a numerical value be given to the property being measured. The devices can program operating cycles at the operator's will, such as compression-relaxation-compression, and the duration of each. At all times the behavior of the material is followed, everything is recorded and analysis of the results lets conclusions be drawn about the property of the texture that is being analyzed.

Furthermore, it is necessary to relate the physical data given by texturometers with the sensations of the consumer. The techniques used are those that are characteristic of *sensory analysis*, which is the set of tests carried out with human panels. Various users, trained or not, depending on what you want to analyze, receive precise instructions to characterize the textures of the product, usually by chewing it. They analyze the different properties and attributes, defining in their opinion the appropriate adjectives for the product, and quantify the property analyzed in comparison with standard products proposed to them. Comparison of the results obtained from the sensory analysis and the texturometers allows a certain food to be characterized by relating quantitatively physics and sensations. These techniques are especially useful for quality control of products being manufactured, assessing acceptance of a new product, or for evaluating the modifications that manufacturers introduce in the products – for example, by substituting the sugar with artificial sweeteners.

There are even more techniques to characterize textures. To eliminate consumer subjectivity, *artificial mouths* have been invented and developed. These are devices that simulate, in a controlled way, the conditions of the human mouth when ingesting a food. They have a mechanical chewing system, the addition of artificial saliva, and the temperature is kept constant. The food is put inside it, and the chewing force and degree to which the product crumbles are measured after a programmed period of time. The results are compared with those from sensory analysis, and qualitative and quantitative conclusions can also be drawn about food textures.

Viscosity, Newtonian and non-Newtonian fluids

On the list of attributes regarding textures is the concept of viscosity, with three descriptors: fluid, thick and viscous. This is a non-scientific, but practical, approach that uses everyday terminology. But there is a more scientific approach. According to science, viscosity is the measure of a fluid's resistance to flow under the action of an external force. There are very simple devices that allow this to be measured, and from it comes a classification for liquids: there are very low viscous liquids, such as alcohol; slightly viscous, such as water; and viscous or very viscous ones, such as oils. All these liquids are called *Newtonian*, for historical reasons. But there are liquids that resist this simple classification such as paints, most cosmetics, and many solid-pasty foods, such as tomato sauce, jams, mayonnaise or ketchup. All of these substances are called *non-Newtonian fluids*, and have quite varied behavior. Liquid chocolate shows non-Newtonian behavior that is similar to ketchup. It needs some initial mixing for it to flow. This behavior is called *pseudoplastic*. When chocolate figures are made from melted chocolate, air bubbles may form in the mass. These bubbles do not disappear on their own, and it is necessary to remove them by stirring the hot mass.

The behavior of fluids can be extremely complex and falls under the domain of *rheology*, a science that studies the flow and deformation of materials. There are fluids that behave like liquids under certain conditions and as solids under others. They are called viscoelastic fluids and are very important in food science. They are characterized by means of *rheometers*, which are specialized texturometers.

Change of texture when making handmade hot chocolate

To make handmade hot chocolate, you have to take a chocolate tablet specifically for hot chocolate – or stone ground chocolate, as it is sometimes called, products 11 and 12 on the tables – and heat it with milk (or water, if you want to prepare Spanish-style chocolate) and stir constantly, preventing it from sticking to the sides and bottom. Chemically, there has been no change: the cocoa butter from the chocolate has melted, mixed with the liquid and formed a thick emulsion. The sugar has dissolved in the water from the milk, and the powder from the cocoa paste has dispersed. Chocolate to make handmade hot chocolate has another component: rice flour or corn flour. With the water, the particles from these flours swell and disperse in the mixture, increasing the viscosity and thickening. But, if we stop stirring, the mass sticks and burns.

The cause of this behavior is the product's high viscosity. When a liquid is heated in a pan that gets heat from the base, its density is reduced and this less-dense liquid will rise to the surface of the pan. The colder liquid that was on the surface will sink to the bottom and heat up. This process is continuous and repetitive: it is how *convection currents* are created, which are very visible in pots and casserole dishes with liquids. But if the liquid mass is very viscous – and more so if it is pseudoplastic, like liquid chocolate – then convection currents are not created. This is because it would require a lot of force to move the mass, and the difference in density between the hot fluid and the cold is not sufficient. The result is that the mass of chocolate is not stirred of its own accord and only heats from the bottom until it burns, while the mass on the surface of the pot remains cold. The way to prevent chocolate from burning is to stir, by hand or with mechanical instruments, the cooking mass and make sure that the sides of the pot do not heat up to high temperatures.

The Maillard reactions

There is a second factor that enhances the effect of "burnt chocolate." These are the Maillard reactions.

In the cooking mass of handmade hot chocolate there is sugar and milk, which even if it is skimmed contains proteins. Whenever there are sugars, flours or different starches in a hot medium, mixed with proteins, Maillard reactions take place. This happens quicker at higher temperatures, usually above 150°C. These reactions are a very complicated set of chemical reactions, still not completely known, which create products having darker colors than the original ones, products with new aromas, which did not exist at the start, and flavors also not present in the initial products. Changes in color, flavor and smell are typical in the kitchen, and these reactions are present in most culinary preparations made at high temperature such as breaded meat, fried fish, bread, cookies, toasted nuts and a long list of others. Caramelization reactions are a bit similar to Maillard reactions, but caramelizations solely come from the sugars present, without proteins. Caramel made from sugar or caramelized onion are examples of this.

When we make handmade hot chocolate, usually the sides and bottom of the pot or casserole dish are much hotter than the mass that is cooking, and it is on the side and bottom of the pot that Maillard reactions between the sugar and milk proteins begin. The viscosity of the mass, along with the Maillard reactions, are responsible for the flavor and smell of burnt hot chocolate so common, unfortunately, in hot cocoa. These Maillard reactions also take place, to some extent, in the roasting of cocoa beans during their harvesting, and are responsible for the changes in color of the roasted beans and for some of the flavors and aromas.

Chapter 4

Taste, smell, color, flavor

The rise in gastronomic culture among the population in recent years is due to globalization, the ease of traveling, fashions and the proliferation of restaurants and shops around the world. Manufacturers have greatly expanded the range of products they offer as well as their sophistication. This has led, among other outcomes, to the development of gastronomic tourism, with many variants and different types of activities. For many years, wine tastings have been popular as a gastronomic tourist activity, and ham, cheese or chocolate tastings are becoming increasingly popular, organized by all kinds of organizations or companies and aimed at the general public. These tastings may have a format similar to the sensory analysis activities described in the previous chapter, or focus only on specific aspects, and usually are about making product lines from the same company or territory known to the public.

There are some basic rules to carrying out a chocolate tasting that has a minimum level of seriousness. The fundamental ones are:

- The chocolates to be tasted will be dark.
- The tasting should be done at room temperature, around 20°C.
- It should go in order from the chocolate with the lowest percentage of cocoa to the one with the highest. When, a few decades ago, the *Lindt* company brought to market its 99% cocoa chocolate, they included instructions for its tasting: it should be consumed after having eaten a little of the 70% and 85% chocolates, so that its bitter taste would not be a surprise right at the start.
- When changing the product, you should rinse your mouth with cold water.

Food tastings do not usually follow the parameters for determining textures in the standards, but rather are generally adapted by the organizers. For example, a commercial tasting can highlight the following parameters: taste, aroma, texture, mouthfeel and fusion. These will be discussed in the following sections.

Gastrophysics and neurogastronomy

Color, flavor, taste, smell and textures are the characteristics of foods that must be evaluated by the consumer. In order to do this, from birth each person has a set of different types of sensors integrated into their organism. When food reaches your mouth, it undergoes all kinds of stimuli and transformations that modify it, causing it to give off aromas and flavors and changing its texture and composition. All these changes are detected by our sensors, which are located, mostly, in the oral cavity and its surroundings.

The influence that the surroundings have on each of our sensors can lead to perceiving the organoleptic sensations in very different ways. The new disciplines of *gastrophysics* and *neurogastronomy* study these influences in detail, which are used by the great chefs to design new ways of savoring their preparations. **Heston Blumenthal**, chef of *The Fat Duck* restaurant, one of the world's most acclaimed, was probably the pioneer and is the person who has taken these ideas the furthest, but he has certainly not been the only one. Advertising also takes advantage of knowledge from these crossover-influences to design new packaging, new products and new sales pitches, and the world of chocolate is no stranger to this. The shape of chocolate tablets, the rounded or sharp edges, the shapes, sizes, colors and flavors of the various chocolate snacks from the major brands are the result of rigorous studies in this field.

As to hearing, it accompanies ingestion and is increasingly being given more importance. The crunchiness of foods, such as in chocolate tablets, is often highly valued by consumers. Sound does not only arrive through the auditory organ, but also through the series of bones in the jaws and teeth, a pathway that is just as important as the external sound source.

Apart from the ear, there are *proprioception* sensors which help modulate the sensations when eating, starting from the movements of the teeth, muscles and tongue. The pressure that teeth put on food and tongue movements are part of our proprioceptive system. This system's receptors are found in the muscles, joints, tendons and bones, and send reports to our unconscious mind and our consciousness about the state of our body.

The colors of chocolate

A mixture's color has to do with the color of its ingredients, but sometimes there are exceptions. Dispersed systems, such as chocolate, have colors that can differ from the colors of their components. Milk, for example, looks white despite the fact that milk fat, which is transparent, is dispersed in the form of droplets, and the continuous phase of milk is water with lactose and proteins, also transparent. The white color of milk is due to the fact that the light that hits it is dispersed because of the microscopic size of the fat droplets, and this dispersion of the light causes the mixture to appear white, even though it does not contain any white coloring or pigment. It is the same phenomenon that happens with cream, beaten egg whites or foams of any type: they all look white even though their components are not.

Chocolate is a mixture of a white component, sugar; a dark one, cocoa powder; and a yellowish one, cocoa butter. If we look more closely at each of these components, we can see that sugar is made up of transparent crystals. But by putting these crystals together, they disperse the light in such a way that the result looks white, as is the case with salt. Cocoa butter looks yellowish due to a similar mechanism: if heated, it becomes a transparent liquid with a slightly yellowish tone. But solidification — that is, the crystallization — of all its fats creates translucent or opaque structures because it is a mixture of microcrystals and the final mixture does not allow light to pass through. The result is the waxy appearance of solid cocoa butter. Lastly, the cocoa powder is dark because it contains brown pigments from the cocoa fruit, which gives color to each particle of the powder. A new form of chocolate, introduced in 2017, is *pink chocolate* or *ruby chocolate*, obtained from certain varieties of cocoa grown under conditions that give it a pink hue.

Milk chocolate is lighter in color than dark chocolate because it contains more sugar and powdered milk. Both components are formed by white-looking particles. White chocolate does not have cocoa and therefore has the color of solid cocoa butter, powdered milk and sugar, and the whole is white-yellowish.

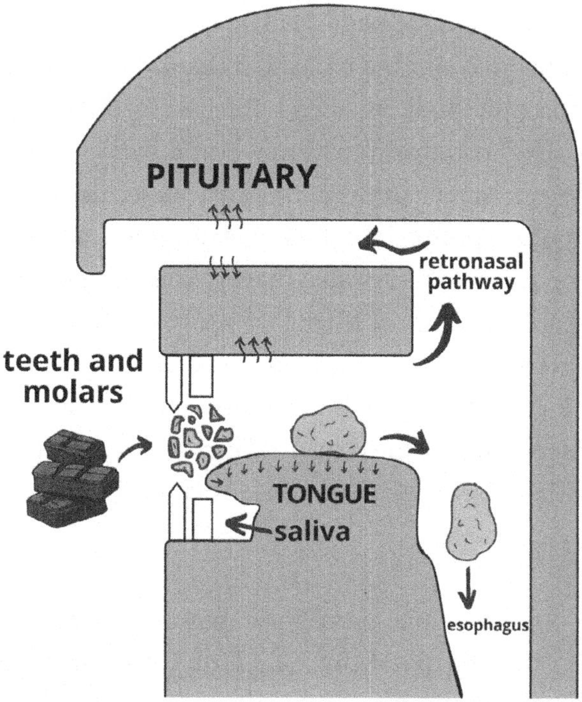

Figure 4. Diagram of a human mouth in the process of chewing and swallowing a food. The food changes its physical state as it is crushed and mixed with saliva. The larynx, parallel to the esophagus, is not shown. Smells reach the pituitary by the orthonasal pathway directly or by the retronasal pathway. Source: prepared by author.

The taste and smell of chocolate

As in other fields, the terminology in the world of sensations is also confusing. In humans, the senses of smell and taste are anatomically differentiated; the pituitary is in the nostrils, and the taste buds in the mucous membranes of the oral cavity and the tongue, which produce independent signals. But the oral cavity and the nasal cavity communicate through the pharynx, and the molecules that smell is sensitive to reach it both by direct inhalation through the nostrils – orthonasal pathway – and by the pharynx – the retronasal pathway. There are molecules that act on both types of receptors, olfactory and gustatory; there are substances that have different molecules, each one acting on a type of receptor. This is why the terminology on sensations is complex: We must distinguish the smell of the aroma and the flavor, and the sensations of the taste. And dictionaries don't help.

There is only partial consensus on the following basic concepts:

- Odor: is the perception picked up by the pituitary gland, by either the *orthonasal* or *retronasal* pathways.
- Taste: the sensations picked up by the taste buds are sometimes called taste. Most commonly, taste is considered to be the combination of the simultaneous perception of the sensations and smell of a substance.

- Aroma: in the food industry, this is any substance that, added to a preparation, changes its taste or smell. In sensory sciences, it is usually synonymous with a pleasant smell, especially if perceived through the retronasal pathway.
- Flavor: is the overall sensation produced by the smell, taste and mouthfeel of a substance.

Generally speaking, the sensations, taste and smell of foods depend on the molecules that make them up and that emerge when chewing, when they interact with the subject's taste buds and pituitary. This is a very important field of research these days, and there is very little consensus among specialists. The complexity of the interactions among the sensory information makes it possible to perceive different tastes and odors depending on the visual environment or the sounds in it. Hence menus – often considered an extravagance of the chef – in which eating dishes is accompanied by headphones that the diner has to wear, with sounds or *ad hoc* music, or other similar practices.

The basic tastes of chocolate are two: sweet, from the sugar – sucrose – it contains, and the bitter taste, which comes from cocoa, and specifically from the polyphenol molecules it has. The sweet taste is always the same because sucrose is a pure chemical product and its taste is independent of its origin. But in the case of cocoa it is different. The bitter taste is general for all cocoa, but it will have more or less intensity, and will be of one type or another, depending on the origin of the cocoa and the roasting it has gone through. In addition, the flavors can be extremely varied, because they combine the taste with the smells that are emitted when it is tasted. The combination of taste, orthonasal smell and retronasal smell can give the chocolate different flavors, depending on the type of cocoa and the preparation it has undergone: hints of spices, fruit, greens, florals, citrus, coffee and others are common. In chocolate tastings, experts educated on these issues claim to find these nuances and many more.

In addition to the flavors chocolate has because it comes from a certain cocoa with specific processing, all variety of products are added to which have also been put in fruits, liqueurs, spices, mint, salt and all kinds of varied external tastes, which create a quite varied gustatory universe.

Texture changes when eating solid chocolate

Before entering our mouth, and supposing it is not a blind tasting, we will have seen the food: color, structure, surface texture, degree of crumbling, etc. All this information that has already entered through our eyes modulates our mind, letting us imagine what it will be like and preparing us for later stages of ingestion. Gastrophysics comes into action before the tasting. The sensations are still not tactile.

Conversely, in the process of biting, chewing and swallowing it, taste and touch are also involved, and a little hearing if the product is crunchy, and that is when all the components of the texture are shown. In fact, the moment of ingesting and swallowing is decisive when it comes to savoring and assessing chocolate, and it is a really complex process.

When it comes to biting and chewing a piece of chocolate – like with any other food – there are several noticeable changes, which depending on the food will be of one kind or another.

- The size of the pieces of food change.
- The temperature changes: it cools down or heats up, as the case may be.
- The food mixes with saliva.
- Aromas are given off.
- The food is mixed with enzymes released by the salivary glands, which can make its composition change.

Temperature. The inside of the oral cavity is at about 37°C. A solid piece of chocolate will melt, therefore, because the polymorph V of the cocoa butter mass, which will be the largest component of any commercial chocolate, melts at 33°C. The feeling of crisp cracking and immediate softening is extremely pleasant, according to tasters and consumers. Melted cocoa butter, although quite viscous, extends over the surface of the tongue and palate, spreading the sugar particles and cocoa powder dispersed inside, and the flavors can be felt throughout the mouth.

Piece size. Chewing the solid pieces of chocolate, at melting temperature, increases the crunchy feeling because the solid surface increases and can melt more quickly throughout the mouth.

Mixed with saliva. The melted product mixes with saliva, which is an aqueous solution. In this mixture, the chocolate's sugar particles dissolve, act on the taste buds and provide the desired sweet taste in the various varieties of chocolate. Likewise, the cocoa powder is dispersed in the saliva and makes the papillae perceive its bitter taste, alternative and complementary to the sweet taste of the sugar.

Release of aromas. The temperature of 37°C allows the mass being chewed to release different aromas present in the chocolate, aromas that at a colder temperature would go unnoticed.

Change of composition. The various enzymes in saliva cause some of the components from the foods to break down, hydrolyze and change their composition, and thus they can change their taste. This happens mainly with sugars and flours, which when degraded normally turn into products that are sweeter than the original ones.

Taste buds

These days, it is considered that there are five basic tastes: sweet, salty, bitter, sour and umami. The first four have always been well known, but umami, discovered at the beginning of the 20th century, has taken many years to be accepted as a basic taste. The existence of other tastes has been proposed, such as fatty, also called oily or *oleogustus*, but for now there is no consensus that it exists as a differentiated taste. Other sensations that are usually considered flavors really are not, such as spicy or burning, which is a stimulus related to pain sensors.

Taste buds are taste receptors located on the tongue, and there are at most about 10,000, which are lost over time as we age. It was asserted – as of 1918 – that the papillae were each specialized in a specific flavor, and that there was a spatial distribution over the tongue, divided into sectors. But currently it has been proven that over the entire tongue there is sensitivity to all tastes.

The pituitary membrane or nasal mucosa has a section specifically intended for the perception of odors. It is located at the top of the mucosa and is called olfactory mucosa. Smell is a chemical sense, sensitive to volatile molecules released into the environment and transported by air. Olfactory cells contain chemoreceptors, where volatile molecules dissolve and produce specific electrical signals that go to the olfactory bulb and the different brain systems.

A sensation that is usually associated with a bitter taste, but is independent and related to texture, is *astringency.* It refers to the alteration of an organic tissue, in this case the oral mucosa, by the action of a substance, and which produces a feeling of dryness. The classic example is wine tannins, but certain chocolates can also have the same effect. It is related to the action of the trigeminal nerve, also responsible for the perception of pain and of spicy flavors.

All the receptors for tastes, flavors and smells, as well as the rest of the signals we perceive, produce electrical and chemical signals that are processed by the nervous system through mechanisms that go far beyond the scope of this book and its author. It is these mechanisms that create the pleasure of eating for us, which is both a physiological and cultural pleasure. We enjoy it in both its facets. Perhaps by eating a Chocolate Paradigm cake?

Chapter 5

Media and chocolate

Once chocolate spread among the nobility and the bourgeoisie, in the form of liquid or handmade hot chocolate, the custom of drinking chocolate at parties and family or social celebrations started to appear in folk literature or novels. A notable example is that of the Barón de Maldá, from Barcelona, who in his lengthy diary "*Cajón de Sastre*," written with extraordinary detail over fifty years beginning in 1769, notes the two cups of chocolate (or more) that he himself drank each day, without exception. There were some religious qualms, especially during Lent, but certain theologians were of the opinion, since it was a liquid, that you could drink it during the days of not eating meat.

Subsequently, the extraordinary worldwide spread of chocolate in solid form, its sweet taste, the pleasure created by consuming it and its (prior) scarcity have turned it into a symbol of pleasure that is in a certain way prohibited or immeasurable. That is why it is present as a main, or secondary, element in many books, tales, films and other media. Let's look at a few examples.

Literature, cinema and chocolate

There are many films and narratives in which chocolate is discussed as a main element of the action.

Roald Dahl, a Welsh author, wrote *Charlie and the Chocolate Factory* in 1964, one of the most widely-read children's books in the world. It is, in fact, a morality tale, where a chocolate maker hides five passes to visit his factory inside some candy bars. The five children who find them are very different, physically as well as in their backgrounds and personalities. During the visit to the factory all of them behave badly, which leads to certain physical trauma and their expulsion from the visit. All of them, that is, except one, Charlie, who was from a very poor family and could only eat one chocolate bar each year. The owner of the factory, Willy Wonka, after discarding the other children, wants to make him his successor to lead the company, but Charlie prefers to continue living with his family. The very detailed descriptions include rivers and seas of liquid chocolate, never-ending chewing gum, and many other wonders. There is a clear moral background, even though the author wished to avoid having a *moral.*

The novel has been considered a work of science fiction for children. Many of the episodes are reminiscent of the life of the author, who often changed countries and schools, and his family endured many difficulties. For example, some of the adults are shown to be bad, and some are good. The author lived through many dramatic times during World War II, such as night landings without visibility or going blind for long periods. He married several times, his wives and children died. He went through moments of deep faith and others of radical atheism. Dahl also wrote novels for adults, often macabre and terrifying, published, among other places, in *Playboy* magazine. All of this made his personality more complex, which is reflected in his books.

The book was made into a film by one of the most eccentric current directors, **Tim Burton**, in 2005. The lead actor, in addition to Charlie (**Freddie Highmore**), is **Johnny Depp**, a regular in Burton's films, who plays the factory owner. The plot of the film is very faithful to the novel, with the aesthetics characteristic of its director.

The novel *Chocolat*, by the British writer **Joanne Harris** (1999), was brought to the cinema in 2000 by the director **Lasse Hallström**. The plot, with a notable dose of magical realism, is also exemplary:

Two outsiders, Vianne Rocher (played in the film by **Juliette Binoche**) and her daughter Anouk (**Victoire Thivisol**), are swept in by the North winds, arriving in a small, very traditional French village in the south where nothing changes over the years. Vianne opens a chocolate shop full of sweets and culinary preparations that awaken passions among the inhabitants. But the parish priest in the village takes a dim view of the shop's presence, and even more so when they prepare a chocolate fair on Easter Sunday. And, as could be expected, the priest ultimately gives in to the pleasures of chocolate.

It is an extraordinarily hedonistic novel and film, where it seems that social problems are solved thanks to chocolate and the protagonist's chocolate shop, which succeed in changing an entire community's way of thinking. It is in line with *Babette's Feast*, where the redemption of a closed community also comes through opening up to a more sophisticated gastronomy.

Probably the first, and most terrifying, tale related to chocolate is *The Chocolate House*, by the **Brothers Grimm**, folklorists and philologists who compiled stories and folk tales from their country, Hesse-Kassel, an independent state at that time that became part of Germany after unification. However, when the Grimms published the story in 1810 it was titled *Hansel and Gretel*, and the witch's house was not made of chocolate because, as noted in Chapter 1, in those days chocolate was a drink and the solid version was not invented until later. In the original story, the house was made of bread, covered in cakes and had windows of sugar. In a way, it resembled the gingerbread houses that are made in places such as *The Fairmont's 22-Foot Gingerbread House*, which is built every year for Christmas in San Francisco.

In later editions of the book, and especially in translations from the late 19th century onwards, chocolate begins to be more prevalent.

In the novel of magical realism *Like Water for Chocolate*, by Mexican writer **Laura Esquivel** (1989), the kitchen and all its surroundings have great importance in the development of the story, but chocolate has barely a tangential role, despite being included in the title. The novel was brought to the screen by **Alfonso Arau** (1992), to the theater, and a television series was also made in 2024.

The novel *Chocolate Blood*, by Czech writer **Radka Denemarková**, is a recent novel from 2025 about the life of George Sand and another writer, Božena Němcová, and the tycoon John D. Rockefeller throughout the 19th century. The chocolate in the title is only metaphorical.

More books in which the term "chocolate" or "cocoa" is in the title: *Cacao* by **Jorge Amado**, from Brazil (1933); *Chocolate à Chuva* by **Alice Vieira**; *Chocolate con veneno* by **Desiderio Vaquerizo**; or *Deseo de chocolate* by **Care Santos**. And many more...

There are other media in which chocolate is present, whether in the title or throughout the narrative. Let's look at some examples.

Fresa y chocolate is a Cuban film from 1994, with political and LGTBI subjects, co-produced by Cuba, Spain and Mexico. The directors were **Tomás Gutiérrez Alea** and **Juan Carlos Tabío** and it is based on a short story by **Senel Paz**, who was also the film's screenwriter. The action takes place in Havana (Cuba) in 1979. David (**Vladimir Cruz**) is a college student who meets Diego (**Jorge Perugorría**), an apolitical and homosexual artist. Miguel, a heterosexual classmate of David's, encourages him to befriend Diego, believing that David will be able to spy on him, as certain politicians consider Diego a threat to the communist cause. In reality, Diego initiates the friendship with only sexual intentions.

The title of the film refers to the famous ice cream parlor in Havana, *Coppelia*: David tells Miguel that he knew Diego was gay because, when he chose ice cream, he chose strawberry even though chocolate was also available. The film received several awards: it was nominated for the Oscar and was shown at the Sundance and Berlin film festivals.

Dame Chocolate was a 150-episode Spanish language soap opera in the United States which aired on Telemundo in 2007. The original script is by Venezuelan writer **Perla Farías**. The lead actors were **Génesis Rodríguez** and **Carlos Ponce**. Making chocolate with a secret ingredient is crucial to the action. It regards the petals of a flower that grows in the village of the leading female character, around a statue of Ek Chuah, the God of Cacao in Mayan culture and one of the most important in this culture. Ek Chuah was also the God of War and of Merchants.

There are many audiovisual products in which the term "chocolate" appears in a metaphorical sense, to refer to the skin color of a certain protagonist or to the natives in the place where the action unfolds. Let's look at two interesting French films.

The film *Chocolat*, by French director **Claire Denis** (1988), is about the decolonization of Cameroon, and the term is used as a synonym for the skin color of the native Cameroonians, a former French colony. The film was nominated for the Cannes Film Festival that same year.

Another French film also originally titled *Chocolat* is from 2016, directed by **Roschdy Zem**. The main actor is **Omar Sy**, and it is a biography of the actor and clown **Rafael Padilla**, born in France to a Mauritanian father and a Senegalese mother and with Black characteristics. He was an innovator in the circus starting in 1886, creating along with his partner **George Foottit** the model for a pair of clowns, the sophisticate and the fool, which has prevailed around the world. This film has been called *Monsieur Chocolat* in many countries, perhaps to distinguish it from the previous one.

In 1980 a Spanish film by **Gil Carretero** entitled *Chocolate* was filmed about the vicissitudes of young delinquents who "*bajan al moro*," that is, go to Morocco to buy hashish. This is because one of the ways in which marijuana or hashish appears is in the form of a solid bar having a more or less intense brown color, which in the world of drug addicts is called "chocolate."

There is also a less obvious relationship between chocolate and music. In addition to the flamenco singer **Antonio Núñez** "*Chocolate*" (1930-2005), who was one of the most notable figures in classical flamenco singing, there are many songs in which chocolate is in the title or in the lyrics. Many are children's songs and from TV shows, but others are from bands in all different eras, which are easy to find online.

Painting and chocolate

The relationship between painting and chocolate seems trivial, because using liquid chocolate to fill sheets of paper is a typical activity of primary school children. But there are visual artists who have turned the use of chocolate into the central element of their work. Italian artist **Giulia Bernardelli**, who uses the pseudonym *Bernulia*, uses coffee and chocolate to create 2D works. These are delicate and detailed paintings, similar to etchings, which need some type of conservation system to keep them alive over time.

Vik Muniz (Rio de Janeiro, 1961) is a renowned photographer who creates objects with all kinds of materials such as waste, pieces of paper or edible materials, among them chocolate. He has made photo series of soccer players or iconic paintings of universal art, which he first creates with chocolate and afterwards photography, destroying the originals and exhibiting only the photographs.

Chocolate sculptures

One of the most remarkable properties of chocolate is its versatility to change from solid to gooey to liquid at nearly room temperature in a controlled way. This makes it a material that lets you quite easily create chocolate sculptures, this being one of the most appreciated arts and one with the greatest creative possibilities.

Amaury Guichon (1991) is a French pastry chef and chocolatier specializing in life-size and even larger sculptures, very colorful and intricate, that include machines —extreme-sized watches, bicycles— and all kinds of animals, comic and video game characters, and full-size real life characters. The artist has a big presence on social media, participates as a judge at different *MasterChef* type contests in various places around the world, and has some books and films about his work. He currently works in Las Vegas (Nevada). **Lluc Crusellas** (1996) is a Catalan chocolatier and pastry chef who we will talk about at the end of the book, just like we already spoke about **Oriol Balaguer** at the beginning.

The creation of chocolate sculptures has its origin in the *mona de Pascua* (Easter cake) of the 20th century. At the Easter festival it was traditional in many Mediterranean regions (Catalonia, Valencia and Murcia) for the godparents of the child being christened — usually the grandparents — to give their godchildren a bread roll or cake with the same number of hard boiled eggs as the child's age, up to twelve. This cake was called *mona*, a term of disputed origin believed to be derived from Arabic or Hebrew. In Catalonia, as of the 20th century, many pastry chefs replaced the hard boiled eggs with chocolate eggs, and later with chocolate figures that reproduce animals or celebrities. It is a rapidly growing tradition.

Chocolate museums

There are more than fifty museums in the world dedicated to chocolate. Some are directly related to manufacturers, such as *Cadbury World* in Birmingham (United Kingdom), the *Lindt Home of Chocolate* in Kilchberg (Switzerland), or the *Valor Chocolate Museum* in La Vila Joiosa (Marina Baja, Alicante, Spain). General aspects of chocolate and cocoa are exhibited in all of them, as well as commercial aspects of the brand.

The city of Vevey (Vaud, Switzerland) has two related museums: the *Alimentarium*, a general food museum that includes chocolate among its products; and the *Henri Nestlé Museum*, which offers a historical overview of all this company's brands, among which chocolate plays a leading role, although these days it has been a little overshadowed by the boom in *Nescafé* and *Nespresso*.

In Barcelona, the *Chocolate Museum* includes the historical aspects and also products; in addition to the museum, it has a workshop where visitors can make chocolate in small quantities, following the philosophy *bean to bar* (from cocoa beans to chocolate bars) in the same facility.

Mediterranean gastronomy with chocolate

Mediterranean gastronomy, and that of Catalonia in particular, have certain characteristics of their own. First, there are various preparations that incorporate meat and fish or seafood in the same dish; and, second, the use of *picada* in many dishes. *Picada* is a mixture of nuts, garlic, oil, herbs, dry bread or crackers, and broth from the dish that is being prepared, all ground together and added at the end of cooking. It is in this *picada* that you can add chocolate, which gives the dish a different flavor and color. Here are recipes for three dishes that use *picada* with chocolate: one made with meat, one with fish, and one made with a vegetable.

Rabbit with chocolate

This recipe for four persons, a summary from the *Corpus de la Cuina Catalana* from the *Institut Català de la Cuina*, is to cook rabbit for four people. If, for whatever reason, you do not want to use rabbit, the same dish can be made with chicken.

First you have to make the *picada*: In a pan, fry two slices of toasted bread, a clove of garlic and the rabbit liver in oil. Once fried, take everything out of the oil and crush it in a mortar with saffron, some roasted almonds and 60 grams of chocolate tablet of 85% cocoa.

Cut the rabbit into pieces and brown it in the same oil. Add a bunch of herbs, a cinnamon stick, two garlic cloves and two chopped onions. When the onions are golden brown, add a glass of white wine, a glass of stale wine or a splash of cognac or other spirit, salt and water.

Towards the end of cooking, add the *picada* dissolved with a little bit of juice from the cooked dish. Remove the bunch of herbs and cinnamon stick. Add salt and pepper to taste and the dish is ready to be served.

Prawns with chocolate

This recipe is a summary from the same book cited above.

For four people, two dozen prawns are sautéed in oil and then set aside. In the same oil, sauté a large chopped onion, two cloves of garlic, a bunch of herbs and a little orange peel. Once the sauté is done, put the prawns back in and add a little water or fish broth.

After about eight minutes, add a *picada* made with a handful of almonds and hazelnuts, 40 grams of chocolate and a little saffron. After two more minutes, the dish is ready to be served.

Cauliflower with chocolate vinaigrette

This recipe is from the chef **Ada Parellada**.

Chop the cauliflower, separating each of its florets, which are baked for 30 minutes at 180 °C with oil, whatever spices the cook prefers and lemon juice.

In a separate bowl, prepare a vinaigrette dressing with melted chocolate diluted with water, Modena vinegar, oil and parsley. This dressing is poured over the cauliflower and the dish is finished off with some pomegranate seeds and *fleur de sel* on top.

Alternatively, you can prepare a bechamel (flour, milk and nutmeg), which is poured over the cauliflower. This is covered with grated cheese and, on top, some grated black chocolate is added. And everything into the oven.

This book began by presenting Oriol Balaguer's *Paradigm* cake, which was named the best dessert in the world in 2001. To keep a certain symmetry, we will close the book mentioning that the Catalan chocolatier **Lluc Crusellas** (Santa Eulàlia de Riuprimer, 1996) won the **World Chocolate Masters 2022** in Paris.

It is an award that has been given since 2005 and is celebrated every three or four years. The next edition will be in 2026. The contest consists of presenting different types of preparations, and the most spectacular test is to create, exclusively with chocolate, a sculpture that fills an entire showcase.

Crusellas designed an elephant ("*The Elephant of Hope*") measuring two meters tall and weighing more than 120 kilos, with a very carefully crafted skin texture. Since 2024, it has been exhibited at the **Chocolate Museum** in Barcelona.

Chapter 6

FAQ about chocolate and nutrition

In much public media, dark chocolate with more than 70% cocoa has been said to be a superfood and is credited with all sorts of nutritional virtues. Many of these so-called virtues are still unproven, but they are spread as if they were already established truths. Below, some of the questions related to chocolate and nutrition have been written down with the intention of noting only proven claims and not speculations.

The Nutri-Score label on chocolate is D or E. Does this mean chocolate is unhealthy?

As discussed above, the Nutri-Score label is an indication of the relationship between sugars and fats, on the one hand, and fiber, vegetables and proteins, on the other. Naturally, in chocolate the first are dominant, and that explains the label. That is, our diet should be based on legumes, fruits, vegetables, and animal and vegetable proteins, and chocolate cannot form the basis of a diet but is only an addition.

Does chocolate contain antioxidants?

Indeed. Several clinical studies suggest that dark chocolate with 70% or more cocoa appears to reduce cholesterol and blood pressure, and even prevent the development of type 2 diabetes. The reason would be the presence in chocolate of flavonoids, antioxidant and anti-inflammatory molecules, also present in other plants. Moreover, it contains large proportions of phosphorus and potassium.

Does chocolate make you fat?

Naturally. In Table 2 on nutritional information, we can see that almost all chocolates have more than 2000 kJ per 100 g. It is a highly caloric food, especially because of the amount of cocoa butter and sugar it contains. Sugar-free chocolates are a little less caloric, about 10%, depending on the product, because they still have cocoa butter but also less sugar.

Does chocolate cause cavities?

All foods that contain significant amounts of sugar – sucrose – can cause cavities to a greater or lesser extent. Sugar-free products avoid this problem.

Is chocolate toxic?

Chocolate contains an alkaloid, *theobromine*, and may contain heavy metals such as *cadmium* and *lead* (see next question). As for theobromine, it is an alkaloid in the family of *caffeine* and *theophylline*, which cocoa seeds contain in some proportion. Like all alkaloids, it is toxic in certain proportions. But, in this respect, there is no need to be alarmed. A lethal dose of chocolate due to the effect of theobromine is equivalent to about 127 tablets of dark chocolate and about 290 of milk chocolate, which contains less cocoa.

Does chocolate contain heavy metals?

Yes, especially *cadmium* Cd and *lead* Pb. The cocoa plant absorbs the ions of these metals from the soil, and that is why cocoa from volcanic areas such as South America have a higher proportion of these metals than those from other sources, such as Africa. There is a disparity in criteria between Europe and the United States as to acceptable intake. As for cadmium, the analyses done on chocolates sold in Spain show cadmium values well below the authorized limits, which are much lower than the values considered dangerous. No limits have been set in the European Union for lead in cocoa. This is an issue that needs to be watched further.

Is chocolate toxic for dogs?

Theobromine metabolizes very slowly in dogs and produces toxic derivatives that accumulate in their nervous system. Thus, chocolate is toxic for them.

Can chocolate go bad?

Chocolate cannot go bad, as its high levels of sugar and fat are not conducive to the growth of microorganisms. Instead of having an expiration date, chocolate has a preferred consumption date, which is usually a couple of years after it is produced. Mold should not be confused with fat bloom, mentioned earlier, which gives a whitish appearance on the chocolate's surface. But this does not mean it is in poor condition.

Is chocolate addictive?

In 1968 the term *chocoholic* was invented, a neologism that is not yet in the dictionary, to refer to people who claim to be addicted to chocolate and cannot do without it. Nevertheless, it is generally accepted that some of the bioactive molecules it contains, such as tryptophan or *phenethylamine*, which do indeed influence behavior, do not cause withdrawal syndrome from chocolate. *Tryptophan* is an amino acid that the brain uses to produce serotonin, an antidepressant substance that lowers anxiety and tension. It is also a precursor to melatonin, a hormone that helps to induce sleep. Because of arguments like these, there are those who talk about chocolate addiction, extrapolating the facts.

Is chocolate an aphrodisiac?

Phenethylamine (PEA), mentioned in the previous question, is produced naturally by the body and, like amphetamines, can cause a certain feeling of euphoria and well-being. In addition, it helps to maintain the levels of dopamine and other neurotransmitters, also related to taste and pleasure. Cocoa and chocolate contain small amounts of PEA, so it helps to increase the sensations mentioned. It also contains interesting amounts of magnesium that contribute to this. Whether it is an aphrodisiac or not depends, I imagine, on the sensitivity of the receptor to let themself be influenced.

Is there nearby, zero miles or fair trade chocolate?

There are non-governmental organizations that strive to market cocoa under fair conditions. For example, *Oxfam Intermón* has agreements with cocoa growers' cooperatives in the Dominican Republic to obtain cocoa with conditions that are sustainable and fair for the farmers. There is the "bean to bar" movement that defends production of the chocolate tablet be made by the artisan from the cocoa bean, with all the processes, except fermentation, done in the same bakery. Many chocolate companies already do it this way. There is also the "tree to bar" movement that can only be done in countries where cocoa trees grow. On the other hand, *zero mile chocolate* cannot exist if there is no cocoa harvesting in the country, but substitute products can exist made with flour from zero kilometer carob seeds instead of cocoa.

Is chocolate a carcinogen?

Some ongoing studies suggest otherwise. Among the flavonoids mentioned in a previous question, there is a molecule, *epicatechin*, which is one of the antioxidants that shows a destructive action on cancer cells while respecting the others. Nevertheless, because excessive chocolate consumption can increase the consumer's obesity and lead to diabetes and various heart diseases, specialists believe that antioxidants in the diet should come from fruits and vegetables, not from chocolate.

Bibliography

Beckett, Stephen T. (2000) "The Science of Chocolate". The Royal Society of Chemistry, London.

McGee, Harold (2004) "On Food and Cooking" Scribner, New York. Especially Chapter 12 "Sugars, Chocolate, and Confectionery"

Spence, Charles (2017) "Gastrophysics: The New Science of Eating" Viking-Penguin, New York

Thoughts

Chocolate won't solve your problems, but an apple won't either.

When no one understands you, chocolate is always going to be there.

Willpower is the ability to break a chocolate bar into four pieces and eat only one.

Titles from the *Quick Immersion Series*:

1. De-Extinctions
2. Populisms
3. Happiness
4. The Science of Cooking
5. Aristotle
6. Jewish Culture
7. Fascism
8. Nonviolence
9. The French Revolution
10. Jazz
11. The Cathedral of Notre Dame of Paris
12. Civil Rights
13. Feminism
14. International Migration
15. The Meaning of Life
16. Religions
17. Modern Architecture and City
18. Anxiety
19. Autism
20. Beauty
21. Chaos Theory
22. Ethics
23. Cinema & Feminism
24. Education
25. Design
26. Chocolate

Made in the USA
Monee, IL
04 November 2026

67379287R00066